What's the story about th.
Big Bang?

For one hundred years, starting in 1915, modern cosmology has been trying to explain how our Universe arrived on the scene without success. These cosmological scientists have been hampered by relying on those 20th century, preliminary mathematical assumptions formulated that have never been supported by replicable physics.

Many theories on the Big Bang event have been promoted, but none of these have convinced the scientific community of their validity, let alone the general public.

Well, technology was not similarly hampered by such early math and proceeded to carry on, making new discoveries, shedding new light on our understanding of physics.

Guided by the study habits of the late Richard Feynman who liked to start from scratch, this author came onto the scene at the best time to take advantage of and to tie together so many vital elements discovered after Feynman's death, all obscured by the above assumptions. And there are a sufficient number of these elements that do not line up to the casual observer.

Telescopes, microscopes, atom/nuclear research, and internet reports of these technological achievements give access to anyone to the factors compiled in this book that finally derive an acceptable explanation, rooted in replicable physics of how the Big Bang — Banged.

Also by Charles Sven

The 21ˢᵗ Century's All New Cosmology

The Big Bang Book

The Big Bang Book
How, Where, & When
Demonstrated

Charles Sven

Center of the Universe Publishing Co., Inc.
Antioch, Illinois

Publisher's Cataloging-in-Publication
(Provided by Quality Books, Inc.)

Sven, Charles J.
    The big bang book : how, where, & when
    demonstrated /
Charles Sven. -- First edition.
    pages cm
    Includes bibliographical references and index.
    LCCN 2015906442
    ISBN 978-0-9670353-1-4

    1. Big bang theory.   I. Title.

QB991.B54S94 2015   523.1'8
                               QBI15-600082

Illustrations and jacket design by Author
Printed on acid-free paper

Dedicated to the memory, methods, and spirit of Richard Feynman, the inspiration for this work

Contents

Chapters

Any study of our Universe must consider the positions,
elements, or current views covered in the following
appendix notes in order to have a credible analysis:

Appendix [notes, references & links] list:
   [All links accessed in May, 2015]

Appendix [notes, references & links] list: Continued

[All **bold** print or [square bracketed text] anywhere in this book is the author's emphasis or notes. (Quotes use Round.)

List of Illustrations

Acronyms used

ANL	Argonne National Laboratory
APOD	Astronomical Picture of the Day
CERN	Conseil Européen pour la Recherche Nucléaire, or European Council for Nuclear Research
CMB & CMBR	Cosmic Microwave Background Radiation
COBE	Cosmic Background Explorer satellite
EMP	Thermonuclear Electromagnetic Pulse
ESA	European Space Agency
FNAL	Fermi National Accelerator Laboratory
GR	General Relativity
GRB	Gamma Ray Burst
HDF	Hubble Deep Field
HST	Hubble Space Telescope
HUDF	Hubble Ultra Deep Field
Lbl	Lawrence Berkeley National Laboratory
Mpc	Megaparsec or 3.2 million light years
NASA	National Aeronautics and Space Administration
SDSS	Sloan Digital Sky Survey
SLAC	Stanford Linear Accelerator Center
SNe Ia	Type Ia Supernova
STScI	Space Telescope Science Institute
WMAP	Wilkinson Microwave Anisotropy Probe satellite

Introduction

> "scientific explanations are always tentative proposals, … but subject to evaluation, modification, or even overturned in light of further evidence."
> Observation noted at:
> philosophypages.com/lg/e15.htm

In all things, it takes time to gather evidence. As more evidence is accumulated, we can expand our understanding that in effect, improves or overturns our prior concepts. However, it takes a long time for such evidence to reach the public, especially if such evidence – attacks long held opinions. This book presents overwhelming, overturning evidence regarding the Big Bang created from nothing.

There are two basic questions that scientist among others are trying to solve – how was our Universe created [Notes etc. in Appendix A.1] & how did life begin? This book will answer the first question by using everyday, replicable physics, and as Richard Feynman [A.2] would, we will start from scratch in order to insure a solid foundation allowing us to describe the How, Where, & When of the Big Bang with clearly apparent or Demonstrable Physics.

The second question is the purview of biologists and philosophers to be taken up by them in their good time.

Currently, scientists have no acceptable theory of what caused the Big Bang. This is the biggest problem confronting all the current theories including the Standard Model/Quantum Field Theory [A. 3] that deals with fundamental subatomic particles and General Relativity.

What cosmology presents today are in reality some very sensational/bizarre [A. 4] concepts, not found in replicable physics, but are based on a number of old, 1915 to 1934, mathematically created assumptions [A. 5], that suggest that a singleton [A. 6] dot popped out of nothing [A. 7] – a condition virtually impossible to imagine – with such a explosive force that supposedly created time, space and matter. [A.8]

The current cosmological acceptance of math as physics, regarding the Big Bang, possesses a barrier to investigation. According to those mathematically created assumptions it is pointless to go back because mathematically nothing existed prior to the Big Bang, so say current views. Fortunately math isn't physics [A. 9] so notes Einstein and Feynman. This book has no barriers to valid scientific or other inquiries.

Mathematics raises its head imperfectly. See what Rowan-Robinson says in his book, *Ripples in the Cosmos*: "Mathematics is a brilliant creation of mathematicians and like all human creations, it contains flaws and limitations. To match mathematics to the universe requires a further kind of brilliance, [giving] insight into physics and into what is known about the universe. This tends to be a patchwork process which most modern writers on physics and cosmology tend to play down."

Mathematical equations without more, only describe quantities – period! Events or qualities studied require a physical demonstration that can be reproduced over and over again. Math cannot replace physics in explaining our Universe, but all the observations and experimental results coming from our marvelous hi-tech environment, available on the web, logically sequenced herein, describe the Big Bang.

Today, the current communication of observational physics as published is very restricted, because most of these discoveries are reported in scientific journals with very limited public distribution, and are filled with obscure scientific terminology and jargon.[A. 10]   Further, trade secrets and proprietary ownership of new discoveries take a long time to reach the public. Such conditions don't simplify or make available new findings for mass consumption.

In spite of such difficult access, information about our world abounds, one just needs to persist in mining these difficult sources now helped by searching the web.   There are in reality no limits to what a persistent individual can find in science [A. 11] about the natural world.

I wanted to know how our Universe was created — 'the something from nothing' cosmological description does nothing to satisfy curiosity – so where to start?   I note that Richard Feynman liked to start from scratch to secure a firm foundation for his work, and basically relied on real observations [replicable physics] as the final judge.

That's my approach; start from scratch.   And, as noted above, the Big Bang supposedly sprang from a singleton dot out of the void, the nothing, the zip, nada, zero; so that's where to start –   at the closest physical replication of a singleton – the atom. [A. 12]   And what a huge volume of information is available buried in the web.   Unconnected knowledge of the working parts inside the atom is a real eye opener – appendix 12 on the atom contains findings, not complicated to follow, but aspects of forces we never learned about in high school.

Bit by bit, with trial and error, man has built up the knowledge base of our Universe. Our expanding technology, from scanning tunneling microscopes [A. 13]   developed in 1981 to orbiting NASA [National Aeronautics and Space Administration] telescopes in space, reported about in the web; provide great observations of how things work that described our ever-changing, constantly expanding view of the world we live in.

We find that technology is very instrumental in unraveling many of the mysteries of today's world – including cosmology. Unfortunately it takes a long time to get that new knowledge into one's daily awareness so necessary to help us release the many preconceived ideas built over time.

Our Universe is very easy to understand when one can show how all the data, derived from new and updated hi-tech telescope and satellite observations, combined with modern experimental findings, and common everyday physics is sequenced and interrelated as presented here in the following chapters. Only replicable 3D physics used in this book.

Three dimensions make up the world that we live in and consequently, only three dimensions are used in this analysis, nor does hi-tech support any more than these three. Other then the concept of time, no one has demonstrated the existence of any dimensions other than three that we live in.

Many students like to ask what if questions. They may be interesting, but there is no end to what ifs. I like to say let's examine the subject at hand with all that we know until we arrive at a question that cannot be answered.

Converting initial assumptions into conclusions too early in the game by reporters tends to make realistic analysis very difficult or almost impossible. Reporters love the sensational, the bizarre, the improbable – such as creation from nothing. We are only concerned with the truth if it's dramatic.

We are living in a great time technologically speaking. Space telescopes currently provide pictures of far flung galaxies with exotic shapes and sizes strewn all over our visible Universe. Hi-tech advances let's us calculate the distances. And it's all available on the web, if you know how to ask the right questions. We know the macro.

The micro is not left out of this picture with the latest in electronic microscopes showing us what atoms look like. Further, great strides tell us much about this magnificent atom. We now know via several studies that virtually all atoms in our Universe were created in the Big Bang era and never break down by themselves. In some very hi-tech labs, we learned how to create a very tiny bit of matter using a huge amount of power. This power related to the atom is almost beyond belief, especially when we consider the chain reaction, the atomic bomb, and the furnace of the sun and stars.

Between the micro and macro finds space filled with wonders. Exploding supernova stars [A. 14] in death throes amaze us. Gamma Ray Bursts [GRBs] [A. 15] the biggest explosions – happening at the edge of space, stagger our imaginations. Our expanding technology allows us to measure these macro items along with micro sensations inundating Earth from all directions referred to as the Cosmic Microwave Background Radiation [CMB or old form CMBR]. [A. 16]

So many mysteries push many impulsive thoughts that without examination, fill our everyday reporters with bizarre concepts, when repeated over and over, drown out rationality, such as space expansion; which is now supposedly accelerating [A. 17], propelled by Dark Energy. [A. 18]

This is the explanation given to explain why the distances between galaxies is increasing — space expansion – a concept not replicated by any physical observation or experiment. Further we do observe that some 7,000 celestial objects including about 100 galaxies [A. 19] are blue-shifted meaning that they are moving towards Earth opposite of what most galaxies are doing. This doesn't make sense if space expansion is the real physical force driving galaxies apart. And further NASA has strong evidence that there are many galaxy collisions [A. 20] taking place – not possible under space expansion.

Along with these thoughts, unsupported by any demonstration of space expansion, is the companion concept of the cosmological principle [A. 21], that 1933 mathematically derived assumption by Milne, that Earth has no special place in our Universe due to our existence in some assumed curved centerless space. [A. 22] We will see that observations prove otherwise. What's left out of this math are other possible solutions to today's mysteries using real physics.

About that Richard Feynman, that very eminent scientist in physics – if he studied cosmology instead of quantum physics, he would be writing this book. He got sidetracked at the start:

From the book, *No Ordinary Genius: The Illustrated Richard Feynman,* "There were, of course, many important fundamental questions, like 'Where does the universe come from?' and so on, that people were working on when I was first learning, and I didn't have a hope – I didn't think to try about that. I had no way to get at them when I was young."

He never did go back to that question; he concentrated on the quantum world.

I have been fortunate enough to be able to pursue this subject at a time and place where technology has excelled and public access is virtually unlimited, although very splintered, via the web, in the area of cosmology and particle physics. And I like to start from scratch using the latest findings available from our expanding technology, just like Richard Feynman.

Great topics that capture our interest have many parts to their makeup and as a consequence one usually finds there are a variety of contributors examining evidence, proposing assumptions and often jumping into conclusions not warranted by in depth observations. What's necessary is a good review of all the elements. As an example:

For one to examine any subject today, such as that old flat earth concept of long ago, many elements, charts and experiments abound that can be assembled into one complete coherent analysis, that demonstrate that the only scientific conclusion is that our Earth is a rotating sphere orbiting a minor star in a not unusually shaped galaxy.

This same is true in our desire to understand cosmology, that study of how our Universe arrived on the scene. There are many seemingly unrelated observations to consider.

During my 18 year search, I found more than enough material to describe conditions that existed prior to the Big Bang and how they contribute to the shape and age of our Universe. NASA's Deep Field observations [A. 23] via the Hubble Space Telescope let's us pinpoint the Big Bang's Epicenter and define the geography and surrounding CMB topography of our Universe while The Planck and Wilkinson Microwave Anisotropy Probe [WMAP] satellites record all the incoming radiation from Space adding to our analysis.

Combining all the above data generates the measuring platform allowing anyone to calculate the age of our Universe, a new expanded age that provides a time frame necessary to explain how gravity can form galaxies.

There are extensive scientific comments and sources in the various appendixes included that help our understanding of how, where, & when the Big Bang took place. The many component parts in this analysis, each of which are observed or demonstrated, have been assembled into one very cohesive model answering the most perplexing questions of how, where, & when was the creation of our Universe, complete with replicable, demonstrable physics.

List of parts:
- Very complicated forces identified in Ageless Atoms
- Power source of atoms and stars described
- Age of atoms determined
- Atoms viewed using new Scanning Tunneling Microscope
- Source of Mathematical Assumptions listed
- Comparison of Math and Physics
- Laboratory conversion of   –   energy into matter
- Atomic Bomb conversion of –   matter into energy
- Recognition of Dark Energy filling all of Space
- NASA's Deep Field's measurements
- Singleton search finds its opposite – the huge CMB
- WMAP & Planck's measuring of – the huge CMB
- Distribution of Supernovas reviewed
- Power of Gamma Ray Bursts – recorded daily
- Quasars & Luminous Red Galaxies mapped
- Published Studies of ThermoNuclear/Atomic forces
- Forces of Nature; On or Off; Detailed
- Thermo ElectroMagnetic Pulse, Turbulence, & Chaos
- New time frame for a proper growth of galaxies detailed
- All Replicable Physics directly based on Observations

This book could have been written by Richard who led the way, in bypassing mathematical assumptions with his 'start from scratch' method of study that can produce the firm foundation so necessary in the development of insight to nature's mysteries. Unfortunately Feynman died in 1988 before half of the vital parts on the above list were discovered.

Most important, along with the availability of personal computers and access to internet reporting of new discoveries, now available to all; I found a key to unlocking the sequence, integrating these parts into one very organized logical tight assembly describing how the Big Bang – Banged.

I hope you enjoy this insight to our Universe.

Charles Sven
Lake Marie
June, 2015

Chapter 1: The How – This study of The Big Bang Explosion & Modern Physics replaces the Singleton with Old Atoms, Dark Energy, and a Stanford Lab Experiment.

For atom notes and reference links, see appendix 12
For singleton notes, etc. see appendix 6
For dark energy appendix 18
For explosion   24

"All things are made of atoms"   Richard P. Feynman
                                                                    Six Easy Pieces

"He [Edwin Hubble] quickly realized what this meant that there must have been an instant in time (now known to take place about 14 billion years ago) when the entire Universe was contained in a single point in space. The Universe must have been born in this single violent event which came to be known as the 'Big Bang.' "
http://science.**nasa**.gov/astrophysics/focus-areas/what-powered-the-big-bang/

" … one thing was fairly certain about the expansion of the Universe. … it has been accelerating … But something was causing it. … It is called dark energy."
http://science.**nasa**.gov/astrophysics/focus-areas/what-is-dark-energy

Here we find NASA emulating Abbé Georges-Henri Lemaître's old mathematical suggestion, that "If the world [Universe] began with a single quantum [that singularity], the notions of space and time would altogether fail to have any meaning at the beginning;" noted in his May 1931 letter to Nature's editor.

These concepts are still held by NASA as noted above. Cosmologists today accepts Lemaître's assumption that no space or time existed before the Big Bang; and that all the atoms in our Universe suddenly made their appearance, where nothing existed before, so condensed that they all fit inside a singleton dot, smaller than this one →   ·

How could this dot   ·   come from nothing?

What do we know for sure? First, the event happened; the Big Bang, that we know, but how did it start?   There are no acceptable theories of how the Big Bang came to be. Scientists rely on the basic mathematical assumptions of the Big Bang springing from a single point where nothing existed before. These assumptions include the concept of expanding space, and the lack of a central point or any special or preferred position by any celestial body; all these assumptions derived from the reworking of Einstein's 1915 math.

Cosmologists really do not know what this singularity [that dot] is, because a singleton popping out of a space-less nothing, really defies scientists, or anyone else's under-standing of physics. Where did that dot come from?

All of our understanding and experiments require a conversion of something into the new whatever. Not only compressing all the atoms in our 100 + billion galaxy filled Universe seems impossible, we cannot even compress a glass of water to half its volume with all our advanced technology.

Figure 1 Compression of Liquids

Construction Manuals:
Chapter 3 Hydraulic and Pneumatic Systems
"For all practical purposes, fluids are incompressible."
http://constructionmanuals.tpub.com/14273/css/14273_96.htm

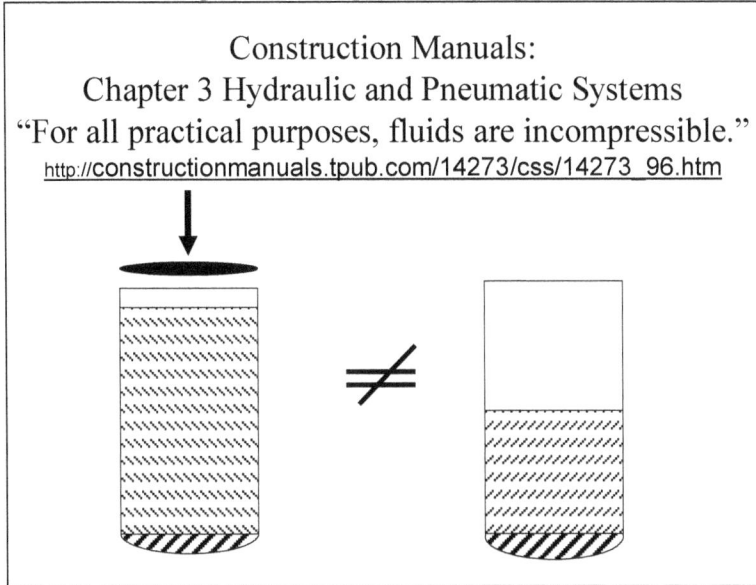

Fortunately for car brakes, hydraulics works, the non-compression of liquids, used every time we step on the brakes.

Regardless of how difficult for us to conceive of – that dot, that singularity   →   ·   we need to start somewhere.

That becomes the major challenge understanding where we should start. Let's start with the closest thing to that single point, closely studied this last century – the atom.

There is much information, buried in the web, about the physical properties of the atom and its subatomic parts, all referred to as the Standard Model / Quantum Theory of Particle Physics or Standard Model for short.

The world of atoms and quantum physics is an arena that one could get lost in the details very easily. However, it is sufficient to just concentrate on the reported major points observed and by doing so we could figure out how they apply to our world and the Big Bang.

The first thing we already know is that the basic atom is composed of electrons and protons. Not as well known is the life span of our atoms. The Theories of the Standard Model predict that the eventually the protons would decay ending the life of the atom – but to the dismay of these theorists, none of their studies, including the latest US/Japan combined Super Kamiokande study, have found any protons decaying, indicating that the minimum life of our amazing atoms is longer then the age of our 13.8+ billion year old Universe.

Figure 2 The US/Japan Kamiokande Study

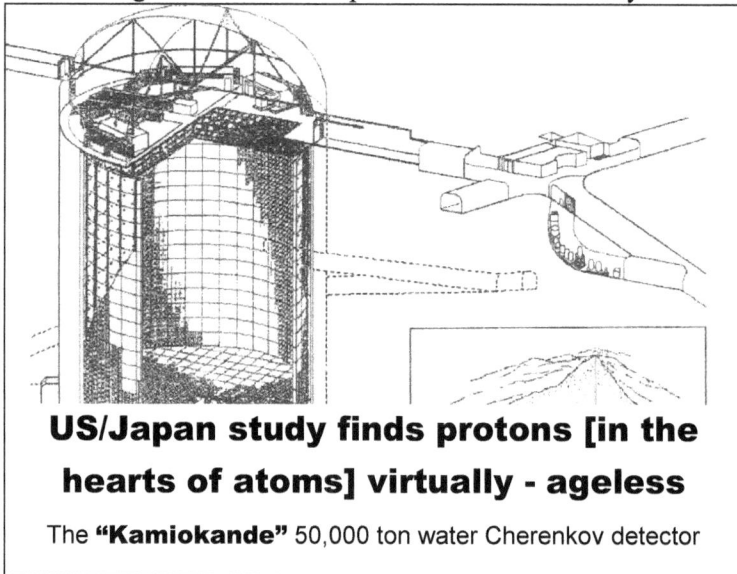

US/Japan study finds protons [in the hearts of atoms] virtually - ageless

The **"Kamiokande"** 50,000 ton water Cherenkov detector

The first dedicated search was undertaken in 1982, by IMB [*Irvine-Michigan-Brookhaven* Collaboration] using an experimental pioneering water Cherenkov technique. We are now at the third generation of this study. The US/Japan Kamiokande [ref. – A12 pg. 122] experiment began in '96 and continues to this day, searching for the atom's proton decay predicted by scientists. No proton decay ever detected.

No proton decay means that the atom has a super long life, currently scientists expect the minimum life of an atom to be 82,000,000,000,000,000,000,000,000,000,000,000 years or in scientific notation: 8.2×10^{34} years which is 8 + 34 decimals.

This means that all the atoms in your [[[the reader's]]] body, in mine and all the atoms existing in our Universe were created in the Big Bang era at least some 13.8 billion years ago. Virtually Ageless. Except for a few very tiny bits created at Stanford labs.

Using the latest improved scanning electron microscope [SEM], we can image the electron orbitals confirming the ball bearing shape of our atoms. The SEM is a microscope that uses electrons instead of light to form an image of atoms seen next, in figures 3 & 4, that look like ball bearings.

Figure 3 – Carbon Atom – a SEM image

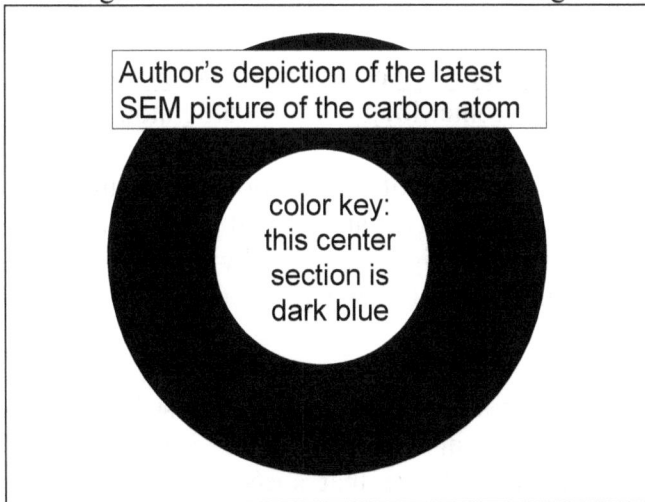

Author's depiction of the latest SEM picture of the carbon atom

color key:
this center
section is
dark blue

http://semisignal.com/?p=1006

Figure 4 A Thin Layer of Atoms

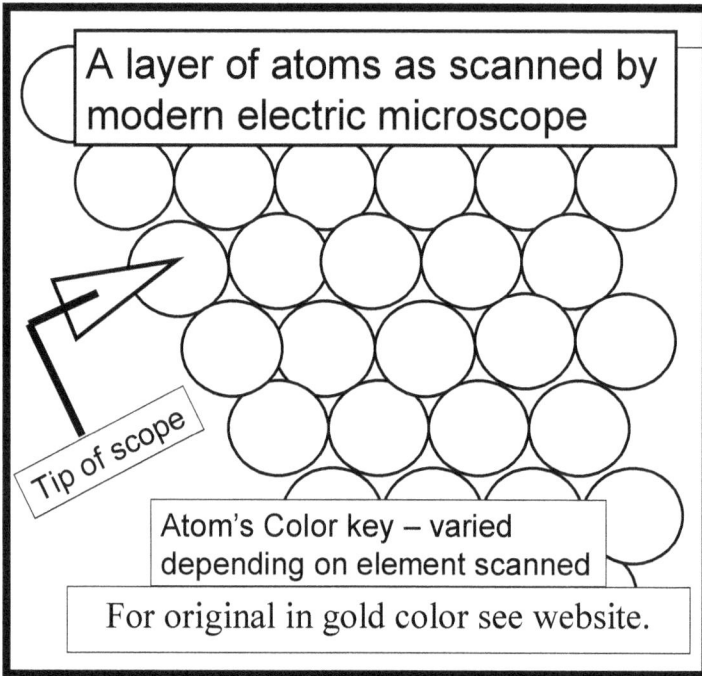

A layer of atoms as scanned by modern electric microscope

Tip of scope

Atom's Color key – varied depending on element scanned

For original in gold color see website.

Credit : Author   For original in color see website:
https://newscenter.**lbl**.gov/2013/05/30/atom-by-atom

This is what a layer of atoms look like using that field-emission microscope. This atom's ball bearing shape provides us with a key to the Big Bang. [[[Why do they look like ball bearings? – answer next page.]]]

To demonstrate an everyday experience, strike a match, and light/photons exit at 186,282 miles per second without acceleration. To achieve that speed requires a transfer of momentum like hitting an eight ball with the cue ball – **instant velocity.**

F. 5 Strike a Match and photons exit with **instant velocity**

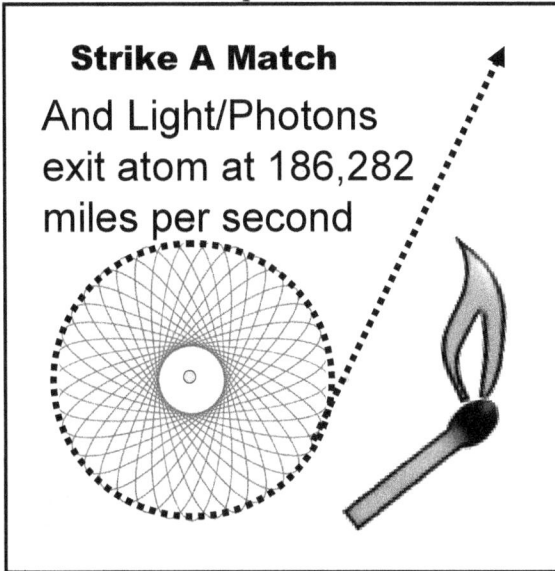

Strike A Match

And Light/Photons
exit atom at 186,282
miles per second

To drive light/photons with **instant velocity**, must mean that something in that atom is already moving at that speed of light, like electrons orbiting its surface, as depicted in figures 3 & 4 above of atoms with their ball bearing shape.

If electrons drive those light/photons at such speed then we must conclude from our observations that the electron is orbiting the atom [like a spirograph path – the existence of a perfect circular path denied by turbulence, chaos, the unending number - π, or the butterfly effect [A31]] at light speed – and is supported by the latest SEM imaging of electron orbitals.

Converting data about atoms into number of orbits requires a little math.   Just divide the distance the electron or light/photon travels in one second – that 186,282 miles by the very tiny circumference of the atom.   The answer   is approximately ≈ **a million trillion orbits per second**. [[[This wavy line math symbol ≈ means approximately.]]] That certainly accounts for the ball bearing shape! This brings us partway to understanding how the Big Bang took place.

Why should a 13+ billion year old atom have the strength to drive a light/photon at the speed of light? We must conclude that atoms require a lot of continuous power since Big Bang's Explosion to drive all the various internal forces operating in the atom to drive that electron for ≈ **a million trillion orbits per second.**

Dark Energy: relationship to atom/electrons – How much power is required to drive those electrons for ≈ **a million trillion orbits per second ?**

Figure 6 Copy of NASA's Composition of our Universe:
Atoms, Dark Matter & 'Dark Energy'

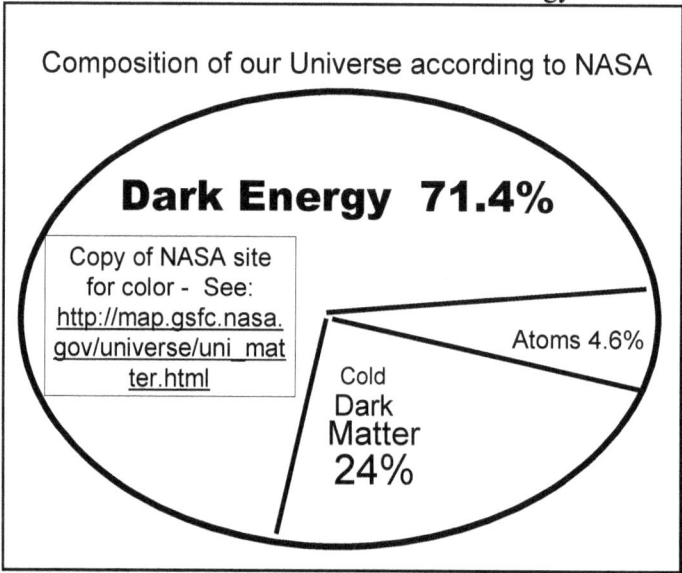

Composition of our Universe according to NASA

Dark Energy 71.4%

Copy of NASA site
for color - See:
http://map.gsfc.nasa.
gov/universe/uni_mat
ter.html

Cold
Dark
Matter
24%

Atoms 4.6%

The Wilkinson Microwave Anisotropy Probe satellite - [WMAP] "determined that the universe is flat"…"total density, we now know the breakdown to be: 4.6% Atoms … 24% Cold Dark Matter…71.4% Dark Energy … [with the] "possibility that the universe contains a bizarre form of matter or energy that is, in effect, gravitationally repulsive" http:// map.gsfc.**nasa**.gov/universe/uni_matter.html

See what the Smithsonian Magazine in April 2010 had to say about 'dark energy' from its article *Dark Energy: The Biggest Mystery in the Universe* By Richard Panek

---

Dark Energy: The Biggest Mystery in the Universe

"We have a complete inventory of the universe," Sean Carroll, a California Institute of Technology cosmologist, has said, "and it makes no sense."

"Scientists have some ideas about what dark matter might be—...—but they have hardly a clue about dark energy." ...

...

"...Michael S. Turner, goes further and ranks dark energy as 'the most profound mystery in all of science'."

[University of Chicago Professor Michael Turner coined this term 'Dark Energy' in 1998, ref. at pg. 151]

http://www.smithsonianmag.com/science-nature/Dark-Energy -The-Biggest-Mystery-in-the-Universe – if link doesn't work search the article title on google – that works

---

We will see how 'dark energy' drives atoms !

It is way beyond our technology to connect a power meter on any one atom. That does not prevent one from getting a good estimate. I believe that dark energy drives the electrons that drive photons, measured by observation of light when we strike a match. Adding to our power study we note that, a tremendous amount of power is released in a chain reaction and in stars. So by inference when we change/disrupt a system one can observe the different effects.

For example, a short in a small electric line or motor creates a small spark. Like a match – small power involved. Knock down an overhead power line and much greater electric flashes or sparks are seen, jumping thru space. By observation one can compare different levels of power involved.

Fig. 7 Spark – Chaotic Electricity – Jumping Thru Space

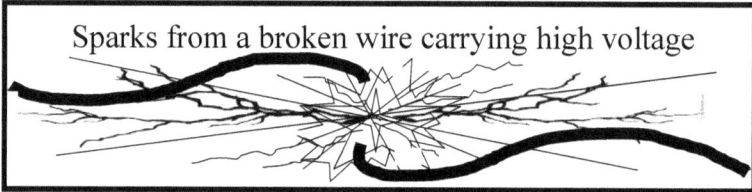

Sparks from a broken wire carrying high voltage

Without putting a meter on some Uranium atoms we can readily see that disrupting the atoms with a chain reaction that a tremendous amount of energy, chaotically destroys the atom and jumps thru space, in effect redirecting 'dark energy'.

Fig.8 Atomic Bomb – Chaotic 'Dark Energy' Observed– compared to a spark – Jumping Thru Space

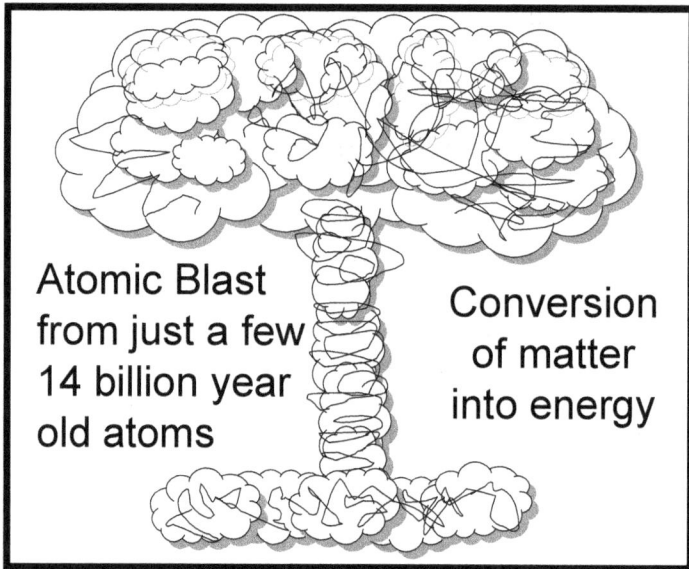

Atomic Blast from just a few 14 billion year old atoms

Conversion of matter into energy

Only a very small amount of Uranium 235 atoms were used in the atomic bomb — just 0.6 tenths of a gram — the number of atoms found in ¼ of a dime. [ref: A12 #10 pg. 128] To just drive light/photons emitted from a match one might not expect a great deal of power involved but —

the power discharged from atoms in a chain reaction is huge, as observed in the chaotic destructive energy coming out of the atomic bomb. This power must be, in effect, the redirecting of dark energy - jumping thru space - that is used to power all the great variety of forces working within the atom.

Dark Energy drives U235 atoms !

Noting the huge powerful forces emerging from an atomic bomb suggest that there cannot be any internal battery found in the atoms but such continuing power must come from an external source – dark energy – jumping thru space.

And now the reverse:

Figure 9 Stanford Lab's Creation of Matter from Energy

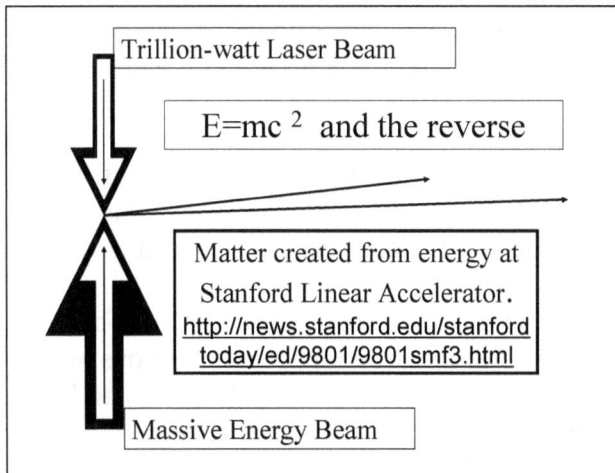

Trillion-watt Laser Beam

$E=mc^2$ and the reverse

Matter created from energy at Stanford Linear Accelerator.
http://news.stanford.edu/stanford today/ed/9801/9801smf3.html

Massive Energy Beam

A study made by the Stanford Linear Accelerator Center in California in '97, converted energy into matter by colliding a trillion-watt green laser beam smashed into an accelerated beam 10 billion times more powerful, creating two very tiny bits of matter, the opposite of the bomb's huge release of energy. Reported in the New York Times – http://www.slac. stanford.edu/exp/e144/nytimes.html

This bit of evidence depicts the tremendous amount of energy needed to convert energy into matter during the Big Bang Explosion – the opposite of the atomic bomb's conversion of the uranium/atomic matter into energy.

It was always theoretically possible and now is physically accomplished: Very high energy is reflected thru atoms via chain reactions, such as the atomic bomb, and now matter converted from the reverse; huge amounts of energy were smashed together at Stanford. This is one of the roles dark energy played in space – creating matter via the Big Bang and now dark energy continuous to power all the many complicated forces operating in the atom.

Dark Energy is keeping all our Universe and body atoms alive, fueling stars, and last, supplied the fuel for the Big Bang. This dark energy took an unlimited amount of time to run, in preexisting space, every available, chaotic, turbulent pattern possible leading up to the Big Bang. This energy must have filled all of space for eons till finally a sufficient amount converged into a small enough space that set off the conversion of energy into matter – just like at Stanford.

The key to all of this was recognizing that some 13+ billion year old atoms found in a match, drive photons at light speed based on web reports made by NASA, US/Japan Kamiokande study, Stanford labs, other equivalent studies, and that match.

Figure 10 The Big Bang's conversion of dark energy into matter shot out into pre-existing space:

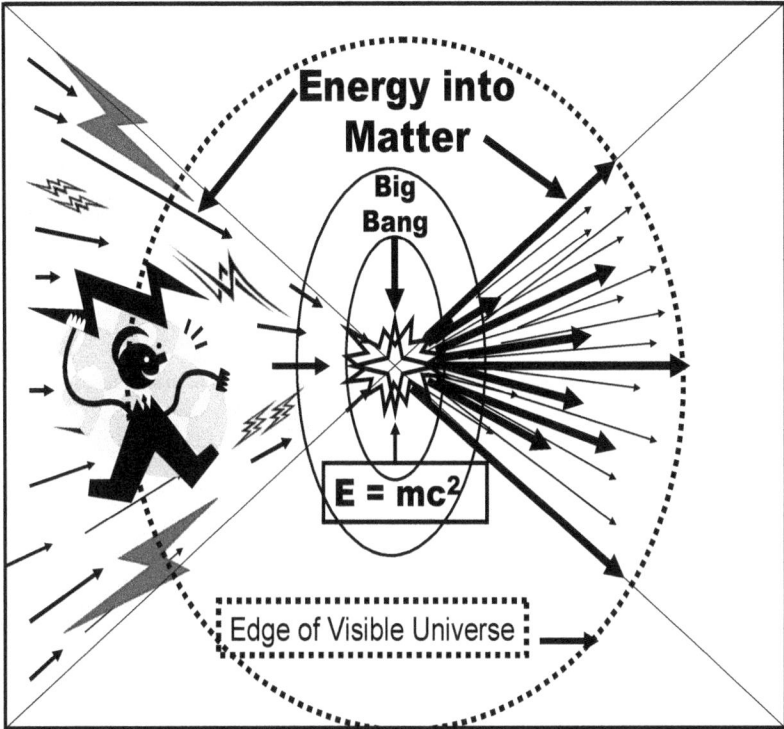

Energy into Matter

Big Bang

$E = mc^2$

Edge of Visible Universe

This concludes the first step – finding the key, that match, that lit up the evidence needed to explain how matter was created in our Universe.

This is How the Big Bang Banged
The How of this Big Bang Explosion – is the conversion of dark energy into atomic matter shot into pre-existing space – the Creation of our Universe.

Chapter 2. The Where: The Triangulation of NASA's North
& South Deep Fields, Cold Spot, and the Cosmic Microwave
Background – all Place Earth in the Center of our Universe.

"One of the main scientific justifications for
building Hubble was to measure the size and age
of the Universe and test theories about its origin…
The first deep fields – Hubble Deep Field North
and South – gave astronomers a peephole to the
ancient Universe for the first time, and caused a
real revolution in modern astronomy."
www.spacetelescope.orgscience/deep_fields/

In chapter one we learned about the Big Bang and 'How'
that conversion of dark energy into matter describes the Big
Bang exploding in preexisting space. That explosive drive
shot out all the hot embers of atomic particles that in time
formed all our galaxies. Like sparks from a skyrocket burst.

This chapter finds us fortunate to have that remarkable
devise called the Hubble Space Telescope at our disposal via
its findings reported on the web — we just have to know what
we are seeing. And to see means organizing the Hubble data
into a NASA '**map**' [figure 13 page 45] that everyone can
evaluate and pinpoint 'Where' the Big Bang took place.

Now Hubble observations will tell us 'Where.' Since '96
NASA pointed its space telescope both north and south
peering as far into space as possible and found what is now
known as the 'deep fields' lying some 13 + billion light years
from Earth. Plotting NASA's deep field observations on a
celestial sphere fig. 11, will provide us with that NASA **map**.

Fig. 11: Celestial Sphere with "X" galaxy plotted:

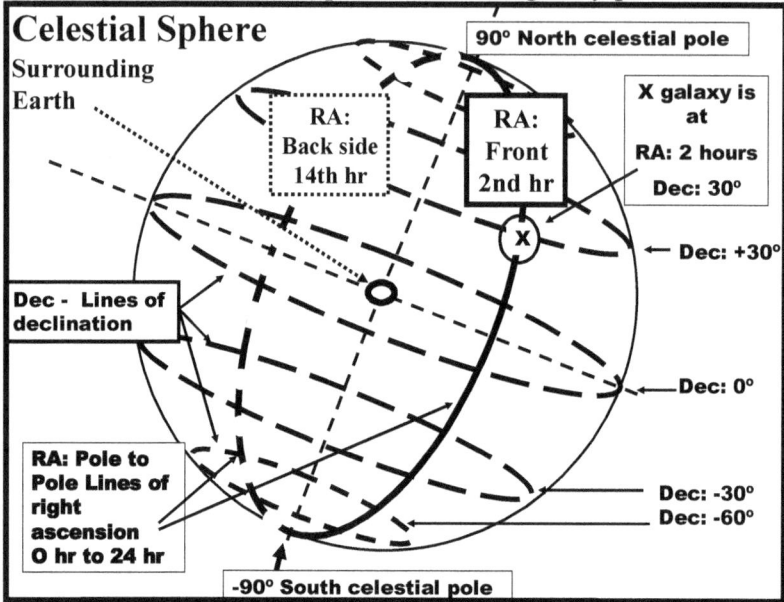

Celestial Sphere
Surrounding
Earth

90° North celestial pole

RA:
Back side
14th hr

RA:
Front
2nd hr

X galaxy is
at
RA: 2 hours
Dec: 30°

X

← Dec: +30°

Dec - Lines of
declination

Dec: 0°

RA: Pole to
Pole Lines of
right
ascension
O hr to 24 hr

Dec: -30°
Dec: -60°

-90° South celestial pole

Fortunately for us, the Hubble Space Telescope's North and South Deep Field findings are reported including celestial coordinates. Details at: *Observations, Data Reduction, and Galaxy Photometry of Hubble Deep Field* [HDF] see: - arXiv:astro-ph/9607174v1 July 31, 1996 - coordinates of the North Deep Field, are given on page 27 as — Right Ascension [RA]: 12hours 36min. 49.4sec. & Declination [Dec]: +62°12'58" We will plot these deep field points in the NASA **map** found on page 45, figure 13.

The Hubble Space Telescope [HST] was carried into orbit by a Space Shuttle in 1990. Its orbit outside the distortion of Earth's atmosphere allows it to take extremely sharp images with almost no background light. As Director of the 500-person Space Telescope Science Institute [STScI], Dr. Robert Williams was allotted by the STScI institute up to 10% of the HST's observation time which was designated as the Director's Discretionary Time.

He chose a most surprising target—nothing at all. That is, a place in space that had no planets or stars or visible galaxies seen with Earth based telescopes. The idea was to see out to the far reaches of space. Its findings were beyond anyone's expectations, gathering light from deep images of the sky showing galaxies to the furthest possible limit, with the greatest possible clarity, from here out to the very edge of our Universe.

How far from Earth is the deep field? The HDF, found many galaxies with redshifts [a 25] as high as six, corresponding to distances of about 12 billion light-years, later increased by the Ultra Deep Field [HUDF] to 13.3 billion — the distance from Earth to the deep fields.

[[[See A.25 - Redshifts — elements give off characteristic spectral lines when gasified and analyzed by spectroscopy. Shifting of these lines to the red acts like the Doppler shift of a moving body. Using the spectrographic analysis of starlight elements, noting spectral lines shifting to the red end of the spectrum indicate distance and velocity from observer.]]]

"That HUDF field contains an estimated 10,000 galaxies." [But] "In vibrant contrast to the image's rich harvest of classic spiral and elliptical galaxies, there is a zoo of oddball galaxies littering the field. Some look like toothpicks; others like links on a bracelet. A few appear to be interacting. Their strange shapes are a far cry from the majestic spiral and elliptical galaxies we see today. These oddball galaxies chronicle a period when the universe was more chaotic. Order and structure were just beginning to emerge." Hubble News Release: STScI-2004-07.

After the great findings of the first deep field, Director Robert Williams pointed the HST south receiving a similar finding. This south target was approximately 12 to 13+

billion light years distant from Earth opposite the north deep field — the galactic coordinates provided by NASA.

Adding to our deep views of the night sky, we need to incorporate the 'cold spot' found in the almost very uniform Cosmic Microwave Background Radiation [CMB] that surrounds our Universe discovered in 1964. NASA identified this lower than average 'cold spot' in 2004, on the CMB — some 13.8 billion light years from Earth. We needed this last coordinate to nail down the Epicenter of the Big Bang and the Epicenter's next door neighbor Earth centrally located in all our observable space identified via **triangulation** as we will see in figure 13 'map'.

Fig. 12: Locating via **Triangulation** of Radio signals:

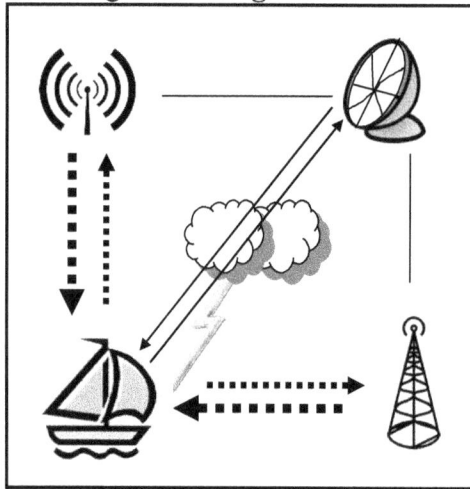

The CMB has fluctuations in the very tiny level of about one part per 100,000 — resembling multicolored sand on a beach. [see cover] The nine year Wilkinson Microwave Anisotropy Probe [WMAP] study found no shifting of these fluctuations eliminating the concept of radiation coming from the Big Bang bouncing around our Universe as the source of this CMB radiation. We will find the true source in chapter 4.

Fig. 13: MAP locating isotropic Earth via **Triangulation** of NASA's data re: the North & South Deep Fields, along with the Cold Spot located in the CMB Ember Radiation Sphere surrounding Earth now 13.8 billion light years in all directions.
[Isotropic Earth detailed– see pages 65 to 69]

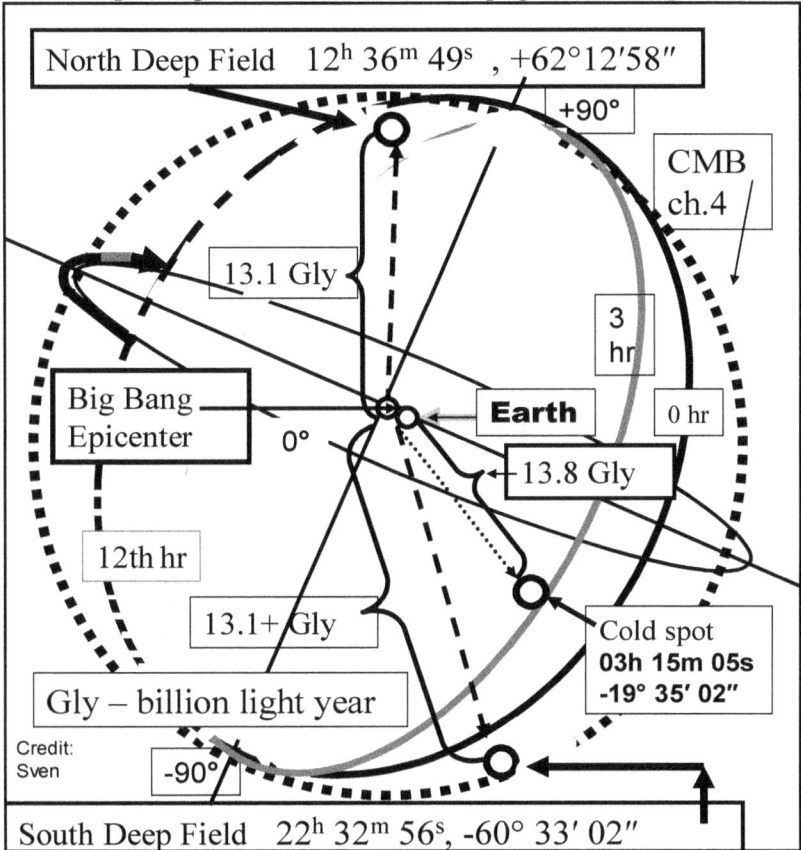

North Deep Field $12^h\ 36^m\ 49^s$, $+62°12'58''$

+90°

CMB ch.4

13.1 Gly

3 hr

Big Bang Epicenter

Earth

0 hr

0°

13.8 Gly

12th hr

13.1+ Gly

Cold spot
03h 15m 05s
-19° 35' 02''

Gly – billion light year

Credit: Sven

-90°

South Deep Field $22^h\ 32^m\ 56^s$, $-60°\ 33'\ 02''$

With all these NASA observations now plotted on above MAP or celestial sphere, we find that by triangulation of this data, we find our Earth positioned virtually right next door to the center of our visible Universe, the second object of my book –

The Where ——
Triangulation of NASA's observable Universe places isotropic Earth next to the Epicenter of the Big Bang.

Chapter 3. The Demonstration – The Big Bang's Tools –
Supernovas, Gamma Ray Bursts, & Dark Energy.
Appendixes 14, 15, & 18

"Supernovas can briefly outshine entire galaxies and radiate more energy than our sun will in its entire lifetime. They're also the primary source of heavy elements in the universe."
http://www.space.com/6638-supernova.html

"Gamma-ray bursts (GRBs) are short-lived bursts of gamma-ray light, the most energetic form of light. Lasting anywhere from a few milliseconds to several minutes, GRBs shine hundreds of times brighter than a typical supernova and about a million trillion times as bright as the Sun."
http://imagine.gsfc.**nasa**.gov/docs/science/know _11/bursts.html

"By observing distant, ancient exploding stars, [type Ia supernovae] physicists and astronomers at the U.S. Department of Energy's Lawrence Berkeley National Laboratory [lbl] and elsewhere have determined [assumed] that the universe is expanding at an accelerating rate -- an observation that implies the existence of a mysterious, self-repelling property of space first proposed by Albert Einstein, which he called the cosmological constant. This extraordinary finding has been named *Science* magazine's "Breakthrough of the Year for 1998."
http://www2.**lbl**.gov/supernova/

"In a letter to Willem de Sitter—discovered in a box of old records in Leiden a few years ago—Einstein wrote, 'This circumstance (of an expanding Universe) irritates me,' and in another letter about the expanding Universe [de Sitter's math in 1917 described expanding space], 'To admit such possibilities seems senseless.' " Robert Jastrow, *God and the Astronomers.* New York: Warner Books, 1978, pg. 17.

"In this picture, although the galaxies are all rushing apart from each other (a fact often expressed in the somewhat misleading statement that the universe is expanding*)
"*It is misleading to say that the universe is expanding, because solar systems and galaxies are not expanding, and space itself is not expanding, The galaxies are rushing apart in the way that any cloud of particles will rush apart once they are set in motion away from each other."   See: Steven Weinberg, *Dreams of a Final Theory,* Pantheon Books, 1994, page 34 and page 34 footnote*.

"The galaxies are not distributed in the universe in a static fashion, like the cities on the surface of Earth; rather, they move apart as though they had been propelled from one and the same point in space by a catastrophic explosion some fifteen billion years ago." See: Henning Genz, *Nothingness*, Perseus books, 1994, page 37.   [like a skyrocket burst]

That the distances between most galaxies[1] is increasing is undisputed. What is required is a scientific demonstration that explains this phenomenon, that of expanding space or an explosion in space. That old mathematical assumption made by Willem de Sitter in 1919 that in a universe without matter – space would expand – has no counterpart in reality.

[1]There are about 100 galaxies whose distances are decreasing a condition not possible if space expands. [references – A. 19]

The only rational explanation for increasing distances between Earth and most other galaxies is that matter races chaotically away from each other when propelled by some explosive force, like the embers of a skyrocket burst.

This chapter will describe an explosion in space using NASA's recordings of very powerful, pulsating Gamma Ray Bursts and how they compare to the scattered distribution of type 1a supernovae [plural also known as type 1a supernovas.]

Prior to the Hubble Space Telescope [HST], we saw these type 1a supernovas fall into a relatively straight line of distances and brightness - the **A** group in figure 14. The farther from earth, the fainter, but with the HST we saw these type 1a supernovas, the **B** group to be off the **A** group scale, a slightly rising curved line based on the average distribution of observed supernovas.

Fig. 14: 1996 Distribution of type 1a Supernovas

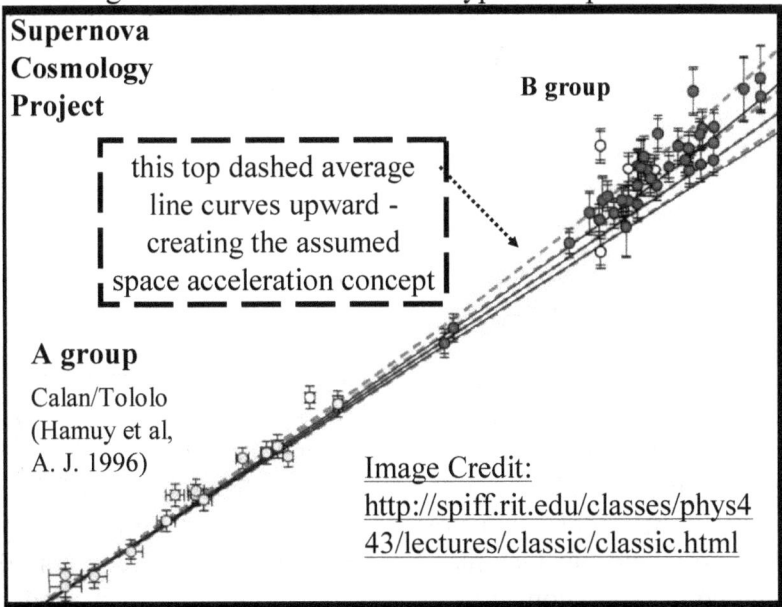

Supernova Cosmology Project

B group

this top dashed average line curves upward - creating the assumed space acceleration concept

A group

Calan/Tololo (Hamuy et al, A. J. 1996)

Image Credit: http://spiff.rit.edu/classes/phys443/lectures/classic/classic.html

The use of averages by the supernova project is misleading – **the scatter plot is reality**! The curved average line is a statistic – not reality!

The assumption of accelerating, expanding space relies on that type 1a Supernovas chart's average of the scattered supernova data, when we need to focus on the **reality**, the scattered distribution. Scattered is what we would expect from a very pulsating Gamma Ray Burst as seen here.

Fig. 15: NASA's graph of a pulsating GRB explosion.

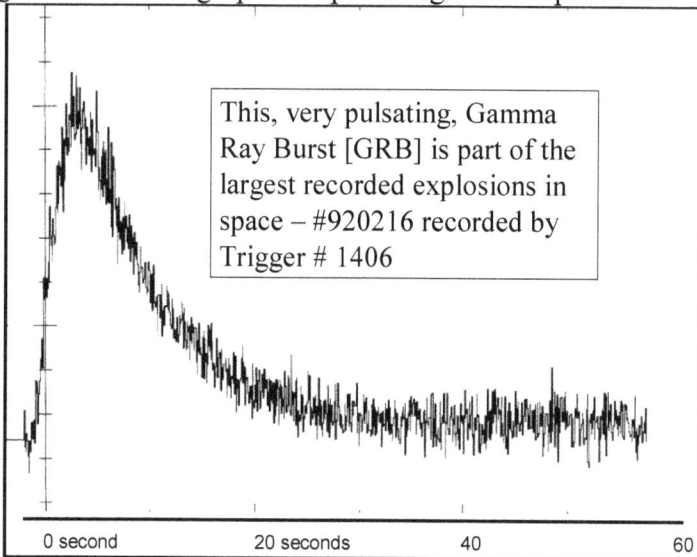

This, very pulsating, Gamma Ray Burst [GRB] is part of the largest recorded explosions in space – #920216 recorded by Trigger # 1406

ftp://legacy.gsfc.**nasa**.gov/FTP/compton/data/batse/trigger/014 01_01600/01406_burst/1406_sum.gif   Credit NASA

This figure 15 Gamma Ray Burst could be used as a model of the Big Bang Explosion. First we need to realize that, although explosions are executed almost instantaneous they all can be understood to require more than just one instant of time [A. 23]. They begin, consume fuel and burn out, in incredibly fast time, and our Gamma Ray Burst shown here requires 60 seconds or so to consume the fuel [dark energy], and explodes about a <u>million trillion</u> times as bright as the Sun, pushing out atomic embers like the continuous ejection of volcanic lava.

Since 1963, NASA discovered these most powerful explosions taking place in our Universe – The Gamma Ray Bursts or GRBs. Currently, orbiting satellites detect on average approximately one GRB per day.

Next we find out that the supernova distribution, figure 14, is scattered; a mirror image of the rapidly declining but very pulsating power graph seen in figure 15. In fact Hubble's 1929 law of Velocity and Distance is also just an average of a scattered distribution of galaxies – the dots in figure 16.

Fig.16   Hubble's 1929 **Average** Line of Velocity & Distance

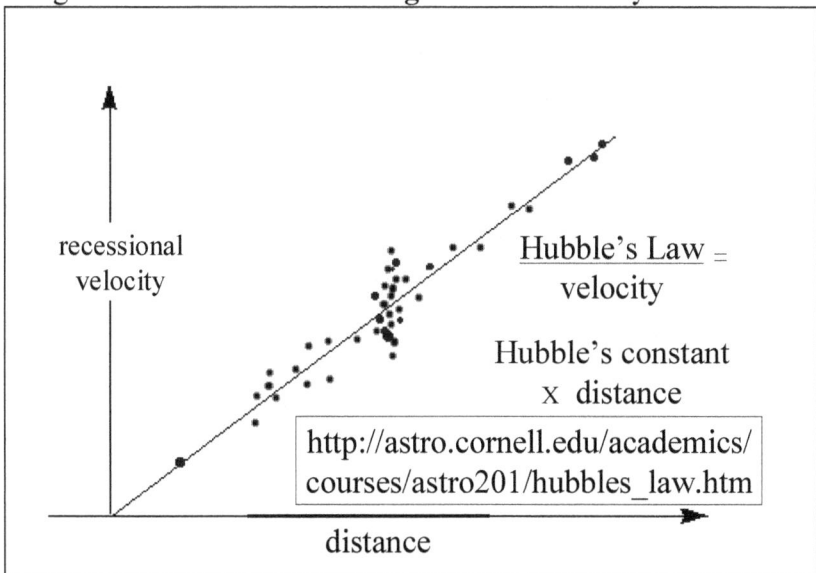

recessional velocity

Hubble's Law = velocity

Hubble's constant x distance

http://astro.cornell.edu/academics/ courses/astro201/hubbles_law.htm

distance

We need to focus on the reality; averages do not convey reality. **Statistics** can be very misleading.

The next chart, figure 17 compares the supernova distribution with sections of the rapidly declining pulsating power curve of the GRB.

Fig. 17: Compare Pulsing GRB graph to Supernova graph.

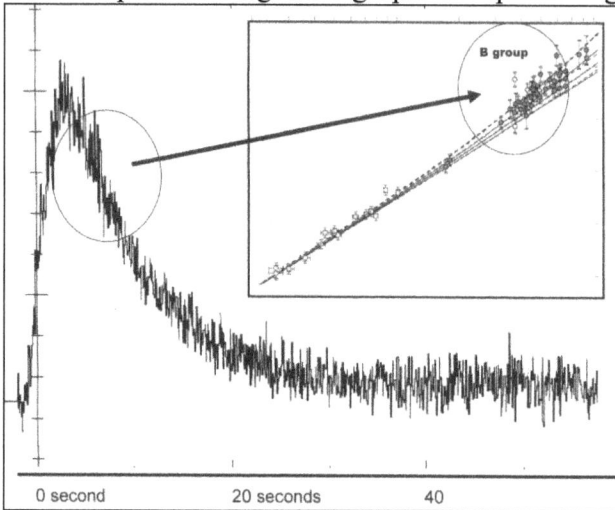

Fig18: Fig 17 GRB graph's relation to all celestial objects

–the CMB ember radiation sphere, Supernovas, & Galaxies

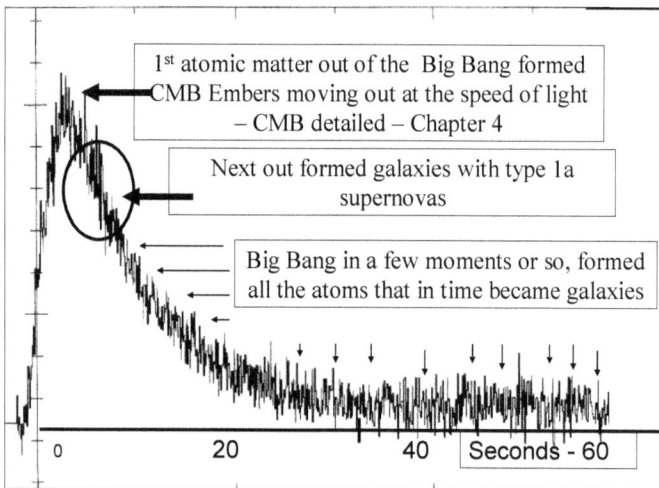

The Big Bang converted the 'dark energy' found in pre-existing space into all the atomic particles that in time formed all the celestial objects seen in the night sky. The 1st out of the explosive conversion, the CMB [Ch.4] shot out at near the speed of light, as the conversion diminished the next in line shot out at slower speeds [like lava out of a volcano] till near the end, the Milky Way & Earth moved at only 0.2% of light speed. – reference figure 21 page 57

Chapter one tells us that the Big Bang was the conversion of dark energy into matter and chapter two found the location, the epicenter of our Universe – so now recognizing that explosions take more than an instant of time; we can add another element – a GRB pulsating explosion going off like a string of firecrackers — shooting embers into space.

That's what we find when we examine this or any GRB graph – a very jagged, pulsating power source indicative of a uneven conversion of fuel into ash – the uneven conversion of dark energy into matter.   —   Like skyrocket embers.

If we can use this figure 18 GRB pulsating power graph as a stand-in for the Big Bang [which probably was 50 million times more powerful in order to create all the atoms-in-galaxies that make up our Universe] we will find that this GRB model describes all we observe coming from – the Big Bang.

So here is The Demonstration.

No expanding or accelerated space required. Just dark energy filling all of space that in the fullness of time, converged into a small enough area with enough compression that set off the Big Bang conversion of energy into matter as seen at Stanford, like a Gamma Ray Burst shooting very hot embers of subatomic particles that in time formed all the celestial objects seen in the night sky including the scattered distribution of supernovas.   [Fig. 10 page 39]

Chapter 4.   The When – Time elapsed since Big Bang using Earth, the Huge CMB sphere & first 3 Chapters.

CMB notes A16

Today, the age of our Universe is measured from the Big Bang event, with the assumption that our Universe began with a singleton exploding out of nothing. Because the speed of light is not instantaneous, when we look out into deep space we are actually looking back in time, supposedly all the way back to the Big Bang.

But —

By looking out as far deep into space as allowed by our technology, cosmologists believe that they are viewing the exploding light of that Big Bang Singleton. However, they didn't find that tiny singleton point; instead **they found the largest structure in our Universe** – the CMB, the Cosmic Microwave Background Radiation from Embers shot out of the Big Bang surrounding Earth.

Fig. 19: Reported by BBC – for color see link below

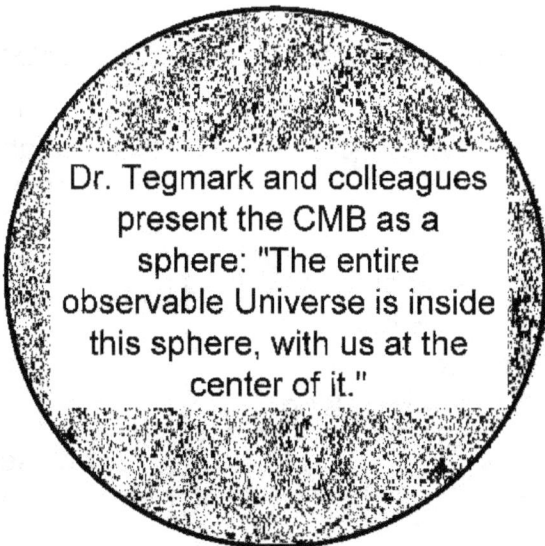

Dr. Tegmark and colleagues present the CMB as a sphere: "The entire observable Universe is inside this sphere, with us at the center of it."

http://news.bbc.co.uk/2/hi/science/nature/2814947.stm

Figure 19: Max Tegmark & NASA data of a radiation coming from everywhere – the CMB. This CMB ember radiation forms a sphere surrounding us as if Earth is in the center of a center slice of this sphere. This sphere is made up of speckles that look like grains of sand on a beach due to outrushing Big Bang embers – see back cover for best perspective and color.

"Does the CMB surround Earth as if we Earthlings are in the Center of our Universe?" … "You're absolutely right that the "true" projection of the [CMB] sky as seen from Earth is akin to the inside of a globe. Astronomers refer to this as the celestial sphere." http://curious.astro.cornell.edu/the-universe/cosmo logy-and-the-big-bang/106-the-universe/cosmology-and-the-big-bang/cos mic-microwave-background/645-why-is-the-wmap-picture-of-the-cmb-an- ellipse-intermediate

[[[This cosmological position – that we did not find light from a tiny singleton dot vs. what is found; the largest structure – that CMB ember sphere surrounding and radiating Earth – is currently not questioned by the scientific community.]]]

Figure 20: How to view the CMB ember radiation

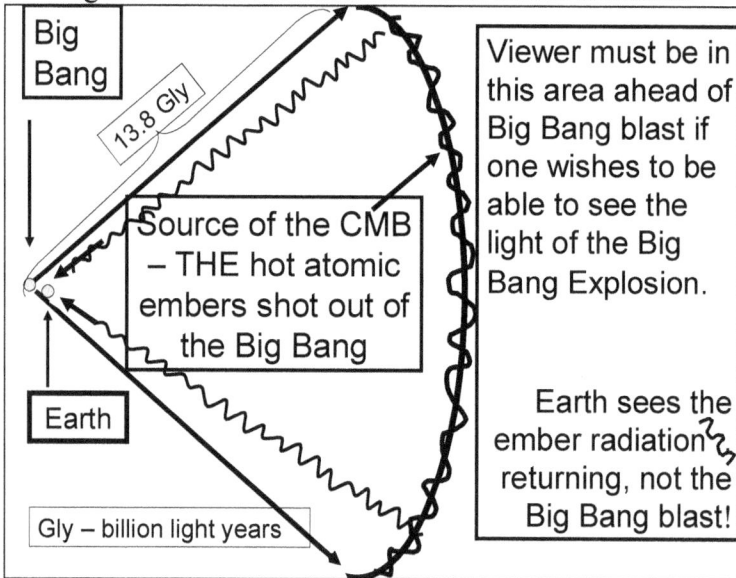

Big Bang

13.8 Gly

Source of the CMB – THE hot atomic embers shot out of the Big Bang

Earth

Gly – billion light years

Viewer must be in this area ahead of Big Bang blast if one wishes to be able to see the light of the Big Bang Explosion.

Earth sees the ember radiation returning, not the Big Bang blast!

And further —

The right hand of cosmology does not pay attention to its left.   According to the current 'space expansion' concept, this assumed math says our Universe is some 41 billion light years in radius   [See A.26] while that Cosmic Microwave Background ember Radiation coming in from all directions [radius] is only ≈ 14 billion light years from Earth.   As a consequence – space is supposedly expanding faster than light speed – so if that is true how could this CMB ember radiation be reflected around to be observed by NASA as a sphere surrounding Earth ?   We should see nothing because in effect the light from the Big Bang cannot reach an edge to reflect back to Earth.   Figure 20.

NASA support of Earth's central location in our Universe:

Fig. 21: NASA's APOD of a very slow Earth 6/15/2014

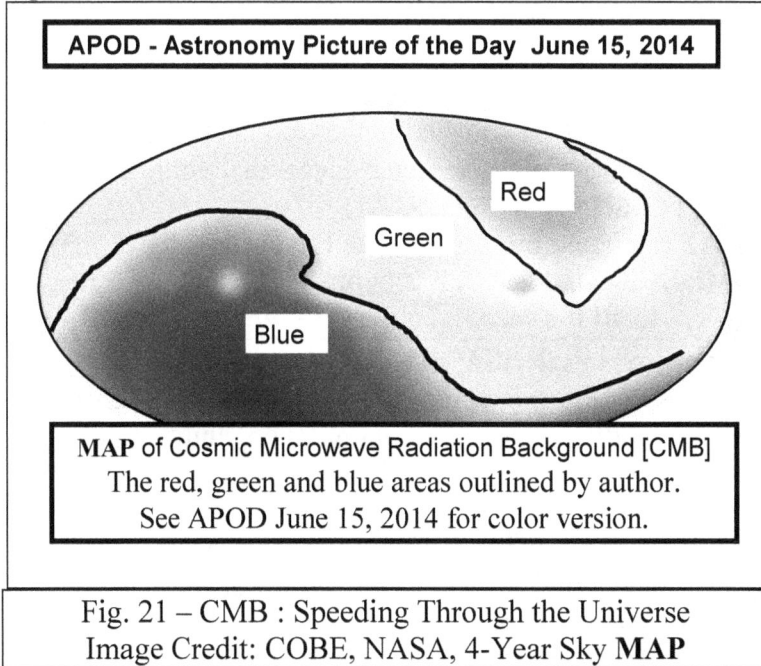

APOD - Astronomy Picture of the Day June 15, 2014

Red

Green

Blue

MAP of Cosmic Microwave Radiation Background [CMB]
The red, green and blue areas outlined by author.
See APOD June 15, 2014 for color version.

Fig. 21 – CMB : Speeding Through the Universe
Image Credit: COBE, NASA, 4-Year Sky **MAP**

Fig 21: This NASA COBE **MAP** indicates that Earth is not at rest. The above [a equal-area "Mollweide Projection."] distorted full sky figure indicates that the Local Group [of galaxies including the Milky Way with Earth in it] moves at about 600 kilometers per second [375 miles/sec.] relative to the CMB.

The above figure 21 [presents the 1993 COBE study] that finds our Earth moving thru space at above explosive velocity when compared to Olympic Standards but when compared to the measuring stick of our Universe we are traveling thru space [like a snail] at a very leisurely 0.2% of light speed, barely exiting the epicenter of the Big Bang. [COBE Reference at arxiv.org/pdf/astro-ph/9312056.]

This NASA COBE Earth velocity study, figure21, when combined with chapter one's recognition that the Big Bang took place in space, with the chapter 3 presenting the power graph of a GRB from high power to last out Earth, and the chapter 2 triangulation of NASA data that puts Earth right next to the Big Bang Epicenter, all support a super slow speed thru space. Also we will find in the next chapter 5, figures 28, 29, & 30 will present telescopic views that confirm our very central position.

All this must be respected as the best evidence that centrally located Earth is moving super slow with respect to all the celestial objects in our Universe.

Note also: this NASA figure 21 is a **distorted full sky** version. Unfortunately all these projections used by NASA, shifts our focus, disabling one's ability to discern the reality of Earth's central location.

NASA uses many very distorting graphics such as their artist concept shown in figure 22 where the CMB does not surround Earth, the opposite of its own findings in figure 23 and as noted by Dr. Tegmark in above figure 19.

Fig. 22 A Distorted NASA Artist Concept of the
CMB 13.8 billion light year distant from Earth

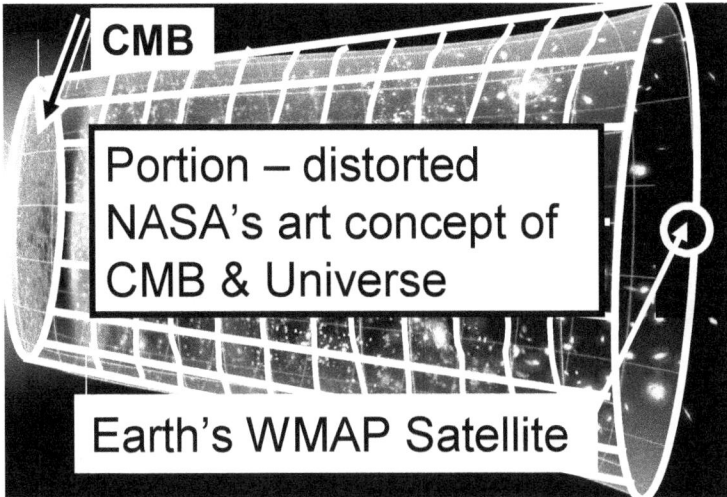

CMB

Portion – distorted
NASA's art concept of
CMB & Universe

Earth's WMAP Satellite

Original MASA artist concept see http://map.gsfc.nasa
.gov/media/060915_CMBTimeline150.jpg

In the attempt to give the full sky picture, NASA only distorts.

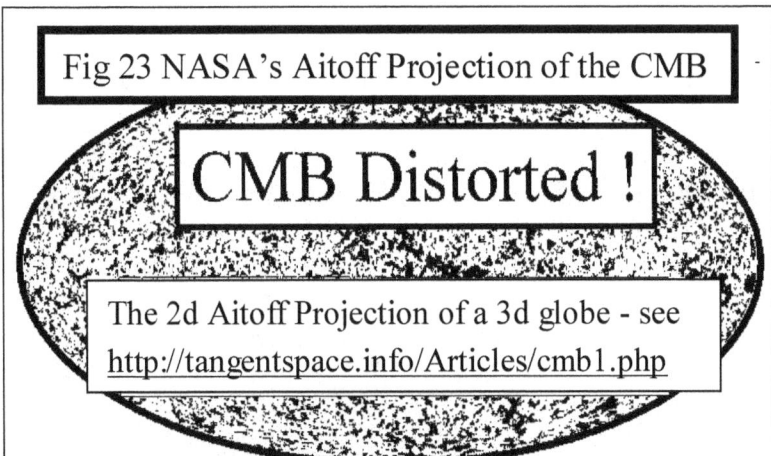

Fig 23 NASA's Aitoff Projection of the CMB

CMB Distorted !

The 2d Aitoff Projection of a 3d globe - see
http://tangentspace.info/Articles/cmb1.php

http://wmap.gsfc.nasa.gov/media/121238/index.html

Fig. 24: NASA's Observable Universe in Undistorted '3D'

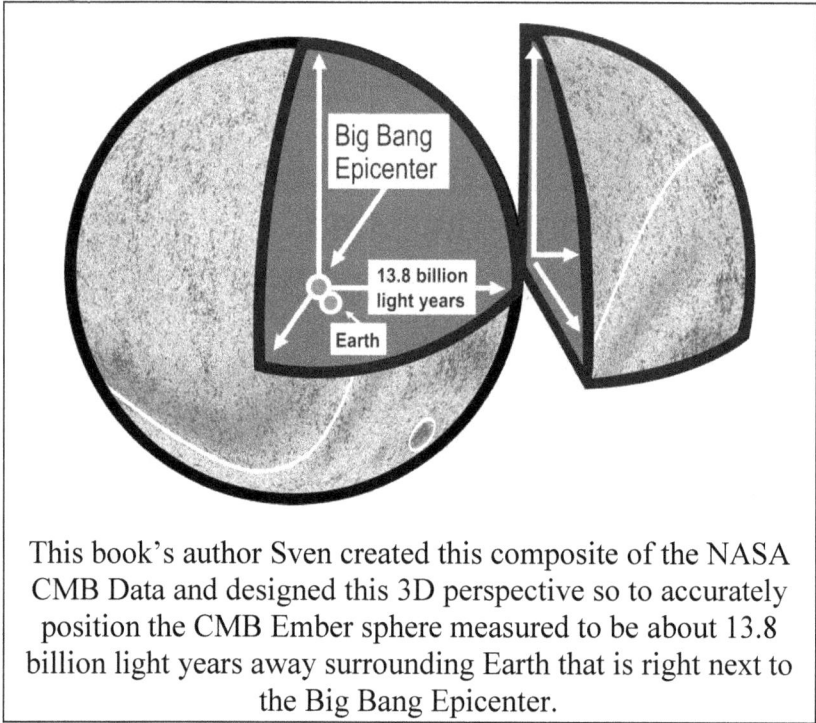

This book's author Sven created this composite of the NASA CMB Data and designed this 3D perspective so to accurately position the CMB Ember sphere measured to be about 13.8 billion light years away surrounding Earth that is right next to the Big Bang Epicenter.

The first thing shot out of the Big Bang, the CMB, the subatomic particles, the now cooling embers, are rushing out at near the speed of light and is the furthest thing that NASA can see, taking some 13.8+ billion years to travel away from Earth; equaling about 13.8 billion light year distance. From that 13.8 billion light year location the cooling CMB ember radiation took a second 13.8 billion years to reach us from that far location — a round trip of some 27.6+ billion years. [[[That's like a operator sending up a sky rocket up some mountain in 1.38 seconds, exploding at the top of a mountain and the flash takes 1.38 seconds to be seen by the operator – or 2.76 seconds after blast off.]]] – see also figure 25:

[[[[[— for another great 3D Earth/CMB graphic — SEE figure 3 at this web site: — http://tangentspace.info/Articles/cmb1.php —]]]]]

60

Fig. 25: Combined Distance of Embers traveling out from Big
Bang Epicenter and returning cooling ember radiation from
the CMB Ember Sphere surrounding Earth received by NASA
[Gly = a billion light year measurement of distance]

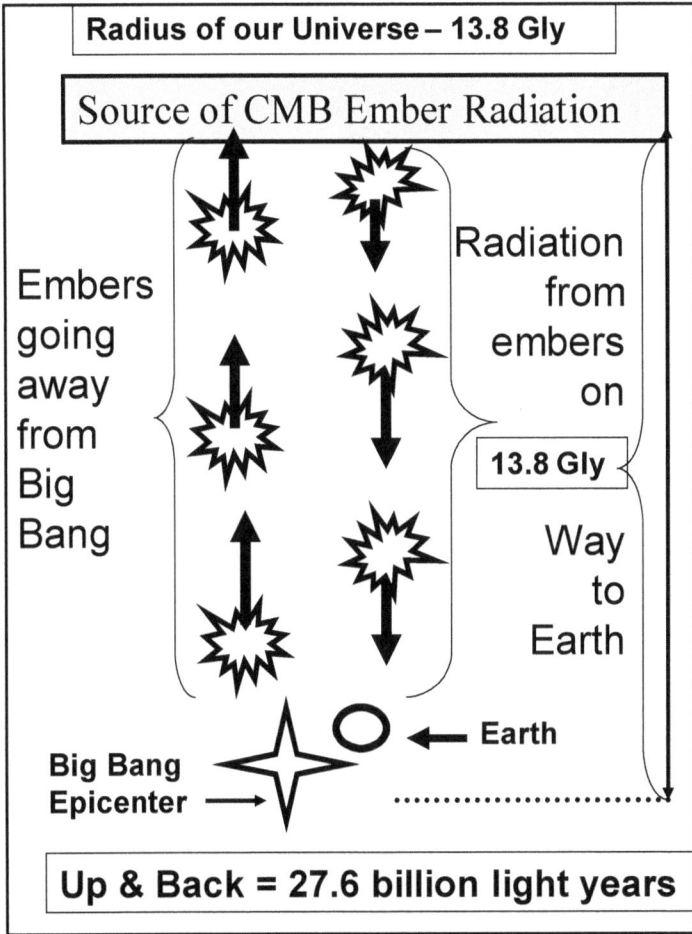

Radius of our Universe – 13.8 Gly

Source of CMB Ember Radiation

Embers
going
away
from
Big
Bang

Radiation
from
embers
on

13.8 Gly

Way
to
Earth

← **Earth**

**Big Bang
Epicenter** →

Up & Back = 27.6 billion light years

The When: **All these observations indicate that the Big
Bang took place approximately 28 billion years ago — the
age of all the atoms in our Universe.**

Chapter 5.   Forces of Nature; Isotropic Earth; Galaxy Formation; Galaxy Age/Distance; and Summary

Forces of Nature – 100 % On or 100% Off
Gravity • Electricity • Magnetism • Light Beams • The Mathematically Assumed Space Expansion & Inflation.
Notes at A17, A27, & A28

On photons – an electromagnetic wave is comprised of photons and that when a light beam is turned on it is always traveling at the local, constant, speed of light – 186,282 miles per second.

This Electromagnetic Wave is either – On or Off

When magnetic lines of force come into existence they also travel at the speed of light another electromagnetic source.

That Electromagnetic Force is either – On or Off

Ah, gravity, it is believed that that force is a force of attraction that exists between any two masses and also travels at the speed of light but is untested. We need a laboratory that has no mass for testing – not available at our current level of technological expertise.

Cannot determine whether gravity is ever – Off

To direct our thoughts to space expansion – current mathematical evaluations say that the size of our Universe is now some 41 to 45 billion light years [A. 26] in radius or around three times the distance to the Cosmic Microwave Background Radiation surrounding Earth indicating that space is growing three times faster than the speed of light. That mathematical assumption of inflation has space growing even faster.

If this mathematically derived, space expansion or inflation is a true force of nature then it must have been turned On with the Big Bang creation of time, space, and matter.

So – in the 1st second after the Big Bang's assumed creation of time, space, and matter, the temperature of our Universe cools to about a trillion degrees allowing the formation of neutrons, neutrinos, protons, electrons, anti-electrons, and photons. But, at the beginning of this 1st second, those assumed mathematical space expansions – are On – and expanding three times or more faster than light preventing the formation of anything but the tiniest component of an atom – maybe a sub-atomic particle smaller than a proton – a quark. Quark references in appendix 12, page 126.

Fig. 26 Atomic Particle or **quark fog** due to Space Expansion

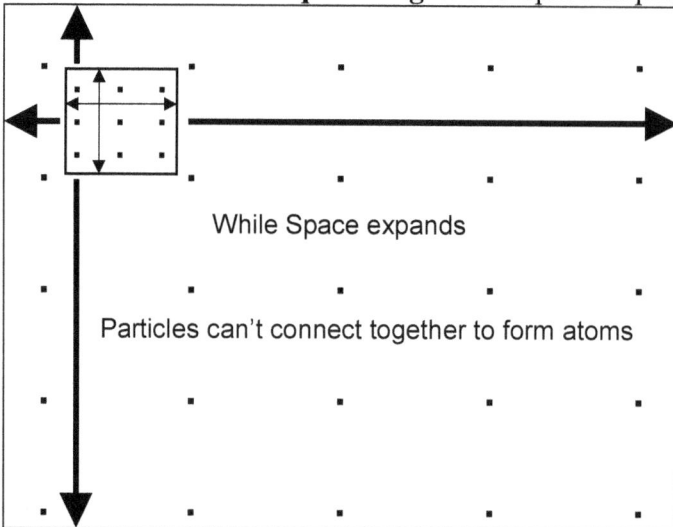

While Space expands

Particles can't connect together to form atoms

In one second, each quark is **186,282 miles apart** if space expands at the speed of light; or 558,846 miles via above mathematical evaluation; and it's even greater with inflation.

Michael Turner doesn't want the cosmological constant, space expansion, or inflation [A.17, 26, 27, 28] to apply at the beginning because if space was growing three times faster than light, such expansion as assumed would be so powerful that the subatomic parts created from the conversion of dark energy into matter would be separated by such space expansion so fast that they could never form a basic atom let alone any celestial objects such as galaxies. All that would exist is **quark fog** – figure 26.

The existence of galaxies proves that The Forces of Nature have no such tool as space expansion.

Astronomical observations place Isotropic Earth in the Center of our Universe.

Figure 27 Isotropic layers surrounding Earth like a Onion

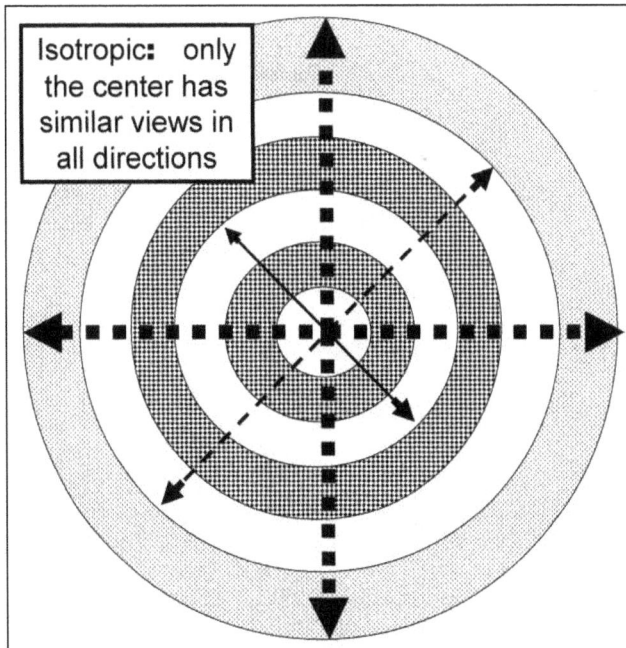

Isotropic: only the center has similar views in all directions

The following NASA observations are based on Earth telescopes. These findings are amazing in that we find basically a center, isotropic, onion-like view of the equal distribution of galaxies in opposite directions going all the way out to the limits of our observable Universe – that CMB ember radiation shell surrounding Earth's central location.

Earth's isotropic views of our Universe – very similar in all directions are only possible from being next to the Big Bang's epicenter in that, the very slow Earth was about the last out of the Big Bang Explosion. The stars in our spiral galaxy obstruct our view, limiting us to pie shaped wedges above and below the Milky Way.

Fig. 28 Isotropic Earth's view of our Universe from the Center of Regular & Luminous Red Galaxies

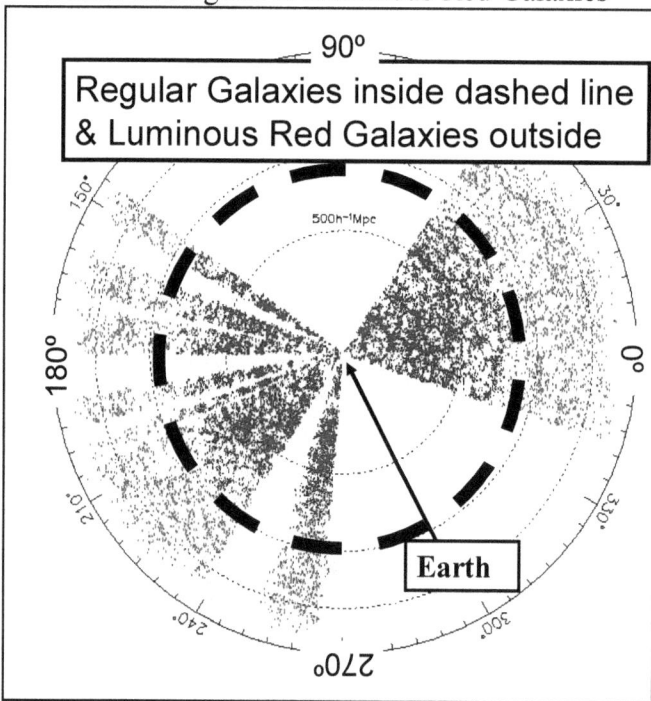

http://www.astro.virginia.edu/class/majewski/astr551/lectures/LECTURE1B/lec1b.html

Figure 29 – Isotropic 2 degree field [2dF] view of Quasars
surrounds Earth each a trillion times brighter than the Sun [A.29]

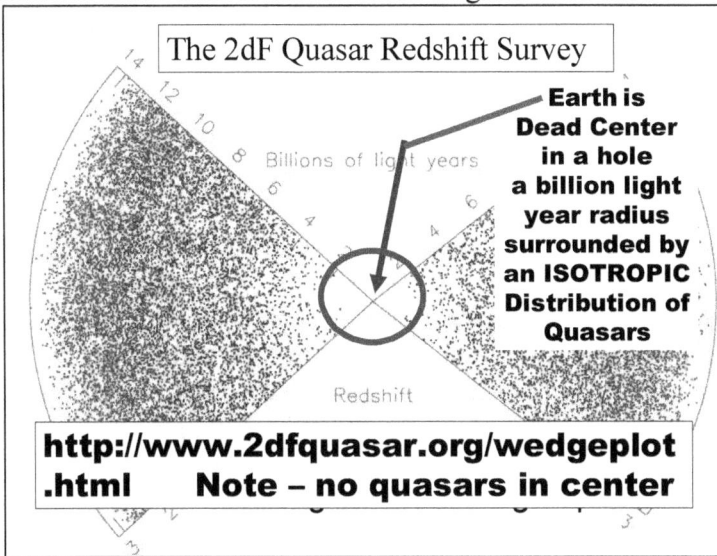

The 2dF Quasar Redshift Survey

Earth is Dead Center in a hole a billion light year radius surrounded by an ISOTROPIC Distribution of Quasars

Billions of light years

Redshift

http://www.2dfquasar.org/wedgeplot .html   Note – no quasars in center

Figure 30: Isotropic Earth in Center of Galaxies & Quasars
Figures 28 & 29 Combined

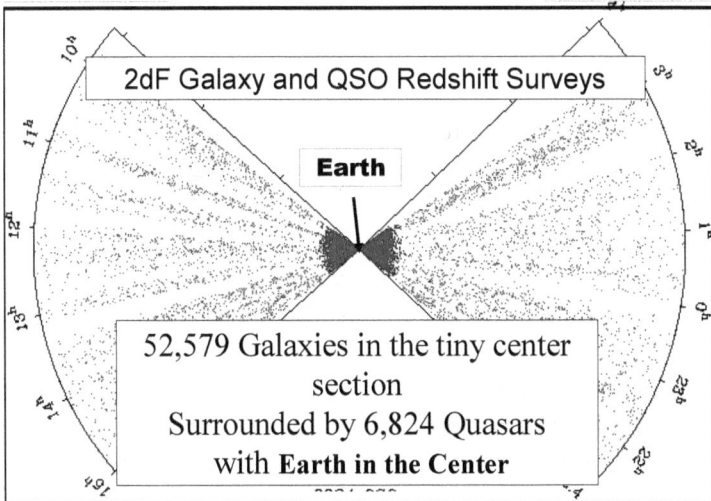

2dF Galaxy and QSO Redshift Surveys

Earth

52,579 Galaxies in the tiny center
section
Surrounded by 6,824 Quasars
with **Earth in the Center**

http://www.2dfquasar.org/Papers/Dunk/wedge.html

As one can plainly see from these observations along with NASA's Deep Field Studies, that very slow, isotropic Earth can only be in one place in our observable Universe — right next to the Big Bang Epicenter.

Isotropic Earth is virtually in the Center of our Universe.

Galaxy Formation & Gravity   Notes etc. at A.30 & 31

"Our galaxy, the Milky Way, is typical: it has hundreds of billions of stars, enough gas and dust to make billions more stars, and at least ten times as much dark matter as all the stars and gas put together. And it's all held together by gravity."

http://science.nasa.gov/astrophysics
/focus-areas/what-are-galaxies/
this **nasa** web site accessed
5/22/2015

Much material has been covered in these first 4 chapters beginning with observational evidence that the Big Bang took place in pre-existing space.

Second we note according to the US/Kamiokande study that all our atoms were created during the Big Bang.

Next, we just described [pages 64-65] how the forces of nature have no such tool as space expansion or inflation preventing gravity etc. from building galaxies.

Gamma Ray Bursts described in chapter 3 graphically depict the beginning, continuation, and end of a super fast pulsating explosion in space, starting very powerfully and diminishes till completion all in a moment or two, like a string of fire crackers, providing the basis of how our galaxies are distributed looking like a bag of marbles dumped – fig.31.

The atomic particles created by the Big Bang was probably accompanied with a chaotic thermo nuclear electro-magnetic pulse force [A.31] seen in all nuclear explosions, acting in concert with gravity on any subatomic particles so created.

Fig. 31 Isotropic Marble drop simulation

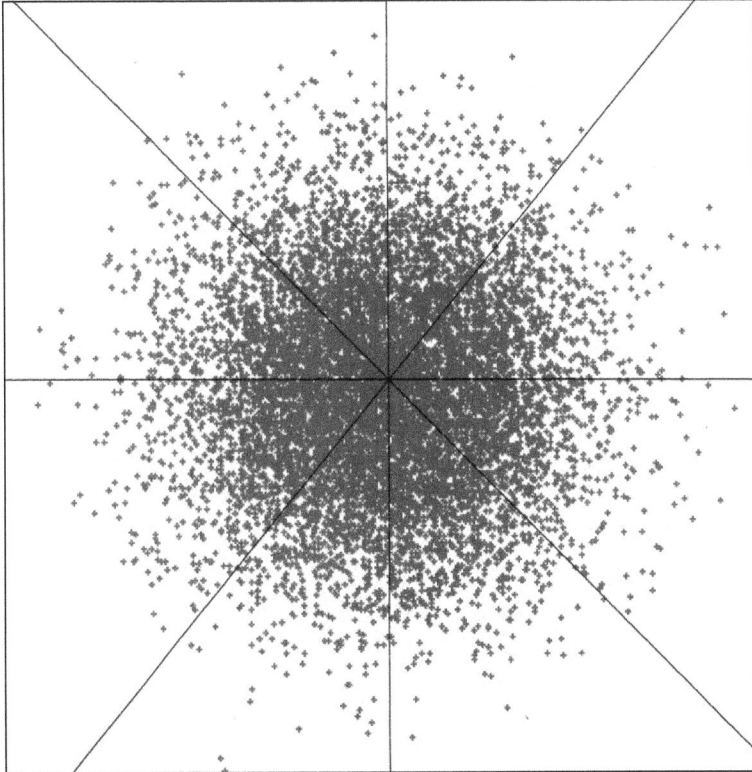

Figure 31 – the isometric aftermath of a bushel of marbles, dropped from off the ground in the center of an open floor – each marble has unlimited degrees of motion, not regimented by space expansion as seen in figure 26 [The Quark Fog]. One could imagine that the distribution of galaxies as seen from our Earth's central position, described in chapter two, could resemble this simulation.

Another factor to be considered includes turbulence that contributes to our inability to predict where a drop of water will be in a stream in a second or two – a turbulence that contributes highly to the 100% degrees of freedom that galaxies exhibit in their myriad number of shapes, sizes, spin, and orientation.   <u>For references on turbulence see page 205.</u>

Current computer simulations of galaxy formations do not match the age of galactic observations because that age is based on the assumption that the youngest galaxies seen just inside of the CMB ember radiation sphere were just created and are supposedly less than a billion years old when in reality those images are 13/14 billion years old, composed of very small, irregularly shaped galaxies moving slower than those fast as light CMB embers. Review again figure 25 page 61.

"I [Jim Peebles] really wonder whether someone will come up with an elegant alternative theory. I may be getting crotchety, but I think people are doing too many big computer simulations and not enough thinking." Michael Lemonick, *The Light at the Edge of the Universe.* Villard, N.Y., 1993, p. 313.

How we need to think of galaxy formation made from Big Bang Embers page 72, figure 32, having 100% degrees of freedom of motion: – they were formed over many billions of years using a variety of forces including gravity, turbulence, a pulsating explosion, and the thermo nuclear electro-magnetic pulse forces. Like embers rushing out from a skyrocket burst.

Looking at all the galaxies in the night sky is like looking at every generation of family pictures on a wall all at once. The oldest generation pictured are teenagers; the next generation in their 20's; followed by next generation in their 30's; continuing up to the present generation photographed when we are in our 90's. There are no pictures younger that those teenagers. We see all these pictures just like seeing all galaxies at once – the old ones nearby all the way out to the youngest just inside the CMB ember radiation sphere.

About galaxies age and distance from Earth
[Gly = a billion light year measurement of distance]
All galaxies are composed of 28 billion year old atoms created in the Big Bang. The furthest that NASA can see is 13.8 Glys in every direction up to and including the 13.8 billion year old CMB ember radiation shell that is moving away from Earth at near light speed. Combined age and distance in light years = 27.6+ or rounded 28 billion years. – The youngest galaxies that NASA sees are slower moving than the almost light speed of the CMB embers.

14 billion year old galaxies are 14 Glys from Earth and are moving away from Earth at near the of light speed. It takes light 14 billion years to reach us – consequently we see the picture of a young 14 billion year old galaxy that took about a 14 billion year delivery time – total elapsed time = 28 billion years after the Big Bang.

28 billion year old galaxy atoms are 0.01 Gly from the Big Bang Epicenter. Virtually no travel time – just 28 billion year old atoms make up local galaxies. We may be 3rd or 4th generation galaxies for all we know, made up of recycled atoms – supernovae star dust.

[[[It takes 21 billion years for atoms to cover a 7 billion light year distance moving out at ⅓ light speed. By adding the 21 billion years moving from Earth plus 7 billion years for light to return from that point = 28 billion years the age of atoms.]]]

The whole Universe is filled with galaxies at every stage of development and their combined age and time to send their image to Earth equals 28 billion years – the age of all the atoms created in the Big Bang.

71

Fig. 32 Isotropically distributed Galaxies propelled by Big Bang in Pre-Existing Space acquiring 100% Degrees of Freedom Generating Unique Orientation, Size, Shape, & Direction

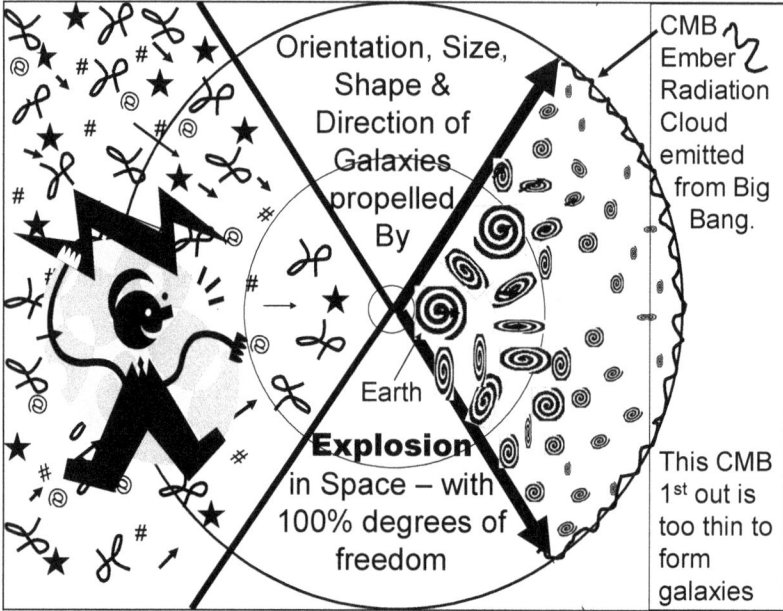

The youngest tiny galaxies just next to, but inside the CMB Ember sphere, moving slower than the CMB Embers, being 2nd out of the Big Bang with less power or push are at least 14 billion years old, built by gravity, etc. allowing them 100% degrees of motion.

28 billion year old, very slow moving, 0.2% of light speed, Earth as part of the Milky Way Galaxy is in the center right next to the cross lines.

Putting it all together

This book would not be possible without the advancements of technology. Calcified mathematical assumptions denied the application of modern replicable physics to all that transpired during the Big Bang era. It was the work habits of Richard Feynman, one of the most highly respected physicists of our times that opened the door to this exploration.

His 'starting from scratch' allowed us to set aside mathematical equations, replacing them with hard evidence — the base to begin this study.

Match/Light is the key – and replication is the Bible. Everything presented here can be observed over and over, verifying the events that led to our existence on planet Earth found virtually in the Center of our observable Universe.

During these 28 billion years, gravity was busy gathering up huge patches of atomic particles that with this long time period available was able to form all the galaxies into the great variety of galactic shapes formed with unlimited 100% degrees of freedom not theoretically possible using space expansion or inflation in our Universe.

All these observations combined allow us to describe the creation of our ≈ 28 billion year old Universe, putting together all the parts listed in the introduction. They also fix very slow moving Earth right next to the Epicenter of our Universe.

Just 'strike a match' thoughts led to this summary:

We Earthlings are located right next to the Big Bang Epicenter of our ≈ 28 billion year old Universe –

created by the conversion of dark energy into matter found in pre-existing space –

starting ≈ 28 billion years ago, gravity, turbulence, and electro magnetic forces created everything that we see.

Appendix 1 How was our Universe created?
No One Knows How the Big Bang Banged
& 'the authority of science'

"we certainty don't know…what banged in the beginning"
"However, the one thing we certainly don't know is what banged in the beginning and how it happened – or even the precise size it had been when it did." Lisa Randall, *Knocking on Heaven's Door, How Physics and Scientific Thinking Illuminate the Universe and the Modern World*, HarperCollins, New York, 2011, page 352.

   This next comment is a great example of what 'the authority of science' says about what is and what is not. Note: This claim says absolutely that there was nothing, no time, and no space before the Big Bang, but on the same page he says there is no physics to explain that! — What a contradiction!

"The second-biggest mistake you can make about space is to imagine an infinite, black, empty space, when, all of a sudden, there is an explosion — the big bang (the 1st mistake is not to imagine at all). The reason this is wrong is that before the big bang there was no space, and before the big bang there was no time. Before the big bang there was nothing. The big bang created space, and the big bang created time. And the big bang created everything that was in it." …
"Not only is this event impossible to imagine, but it is also impossible to describe. We do not have the physics to correctly explain the first instants of our universe — but we can try, hopefully doing better in the future." Richard Hammond, PhD, [Hammond is an Adjunct Professor at the University of North Carolina at Chapel Hill and works for the Army Research Office as a theoretical physicist.] *The Unknown Universe: The Origin of the Universe, Quantum Gravity, Wormholes, and Other Things Science Still Can't*

Explain, New Page Books, A division of the Career Press, Inc., Franklin Lakes, NJ, 2008, pages 213 & 214.

"the big bang theory says nothing about what banged"

"Then there are profound and irresistible questions about the big bang. Why did it happen? What preceded it? Was it a unique event? MIT professor Alan Guth[*] has commented: 'the big bang theory says nothing about what banged, why it banged, or what happened before it banged'." Chris Impey, *How It Began: A Time-Traveler's Guide To The Universe*, W. W. Norton & Company, New York, 2012, page 314.

[*] [Alan Guth comments published in the Fall, 1997 SLAC publication *the Beam Line* article titled: *Was the Cosmic Inflation the 'Bang of the Big Bang?'*]

"If there were an instant, at a 'big bang,' when our universe started expanding, it is not in the cosmology as now accepted, because no one has thought of a way to adduce objective physical evidence that such an event really happened." P.J.E. Peebles, *Principles of Physical Cosmology*, Princeton University Press, Princeton, 1993, page 6.

"How did we get here? … There are many questions associated with the creation and evolution of the major constituents of the cosmos." NASA Official: Kristen Erickson; Last Updated: Nov. 7, 2014. http://science.nasa. gov/astrophysics/big-questions/how-did-universe-originate- and-evolve-produce-galaxies-stars-and-planets-we-see-today/

"the Big Bang, … is a complete mystery"

Sean Carroll: *Why Does the Universe Look the Way it Does:* "In cosmology, we have the Big Bang, which is a complete mystery. How did the universe begin?" John Brockman,

Editor, *The Universe: Leading Scientists Explore the Origin, Mysteries, and Future of the Cosmos*, Harper Perennial, New York, 2014, page 99.

" In particular, what happened at the infinitely dense point, or singularity, from which the Big Bang sprung?" …"In the current best theories that we have, we know that we don't know," says Sean Carroll, a theoretical physicist at Caltech." http://www.pbs.org/wgbh/nova/blogs/physics/2013/09/was-our-universe-born-in-an-extra-dimensional-black-hole/

"Scientist: Canterbury University physics associate professor David Wiltshire. — "We do not know the answer to the question of where all the mass-energy in the universe came from, ultimately. We are still working on understanding that." http://www.stuff.co.nz/science/7313951/Ask-a-Scientist-What -caused-the-Big-Bang

"The universe began, scientists believe, with every speck of its energy jammed into a very tiny point. This extremely dense point exploded with unimaginable force, creating matter and propelling it outward to make the billions of galaxies of our vast universe. Astrophysicists dubbed this titanic explosion the Big Bang." http://www.exploratorium.edu/origins/**cern**/ideas /bang.html

Questions and Answers at Harvard web site: Was the Big Bang the origin of the universe?   Accessed 3/28/2015.
"The Big Bang scenario simply assumes that space, time, and energy already existed. But it tells us nothing about where they came from - or why the universe was born hot and dense to begin with." https://www.cfa.harvard.edu/seuforum/faq.htm#m12

What happened before the Big Bang? Dr. Dave Goldberg,
Ask a physicist 2/2/2012: "… not only don't we know what happened before the Big Bang, we don't even know what

happened in the instant immediately following the big Bang."
http://io9.com/5881330/what-happened-before-the-big-bang

Matter and Anti-Matter in the Universe – If the Big Bang was pure energy, what made the matter? (Submitted July 10, 2009) "Now, where the radiation that was converted into the hydrogen came from and what came "before" the Big Bang, these are questions astronomers don't know the answer to right now and may never know." http://imagine.gsfc.nasa.gov/ask_astro/cosmology.html

"… we will assume that equal numbers of positrons and electrons were present at creation and look to the laws of physics to tell us how there came to be more of one than the other at a later date." James S. Trefil, *The Moment of Creation: Big Bang Physics From Before the First Millisecond to the Present Universe*, Charles Scribner's Sons, New York, 1983, page 33.

"Was there an era before our own, out of which our current universe was born? …We do not know." Jenny Volvovski, Julia Rothman, and Matt Lamothe, *The Where, The Why, and The How: 75 artists Illustrate Wondrous Mysteries of Science*, Chronicle Books, San Francisco, 2012, page 12.

"However, as the mathematical physicist Stephen Hawking points out, a proper formulation of this concept of the beginning of time, as well as that of space, must await a quantum theory of gravity, should it be forthcoming." Joseph Silk, *On the Shores of the Unknown: A Short History of the Universe*, Cambridge University Press, New York, 2005, pg 3.

"But both the beginning and the end of time would be places where the equations of general relativity could not be defined. Thus the theory could not predict what should emerge from the big bang. … we don't yet have a complete understanding of the origin of the universe." Stephen Hawking, *The Universe in a Nutshell*, Bantam Books, New York, 2001, p.24.

"Indeed, we don't even know for sure that our Universe really had a beginning at all, as opposed to spending an eternity doing something we don't understand prior to Big Bang nucleosynthesis." See Max Tegmark, *Our Mathematical Universe*, Alfred A. Knopf, New York, 2014, page 65.

"We may find a nearby wormhole or naked singularity and exploit it as a time machine. We may develop the technology to be able to travel to other universes. We will see for ourselves how the universe began, and no doubt we will come to understand the nature of dark energy and dark matter, the two great enigmas of our epoch." Joseph Silk, *Horizons of Cosmology: Exploring Worlds Seen and Unseen*, Templeton Press, 2009, page 186.

"we have no answer to what happened *before* the explosion"
"Since mathematics and physics don't work at the singularity – and we have no answer to what happened *before* the explosion, if that is when space and *time* began – theories of the beginning of the universe typically start a tiny fraction of a second after the big bang." Amir D. Aczel, *God's Equation: Einstein, Relativity, and the Expanding Universe*, Four Walls Eight Windows, New York, 1999, page 171.

" creation of the universe remains unexplained"
"The creation of the universe remains unexplained by any force, field, power, potency, influence, or instrumentality known to physics—or to man." See: David Berlinski, *Was There a Big Bang?* Commentary, February 1998, page 16.

A. 2 Richard Feynman: His 'start from scratch' problem solving approach & comments on photons

Richard Feynman [1918 – 1988] – pronounced 'Fine – man'

"starting from scratch, solving it in his own way"
"Feynman needed to fully understand every problem he encountered by starting from scratch, solving it in his own way and often in several different ways." Lawrence M. Krause, *Quantum Man; Richard Feynman's Life in Science,* W. W. Norton & Co., 2012, page 7.

"Although Feynman was never seriously interested in cosmology, he did have an interest in general relativity and a good knowledge of the situation in cosmology." See: Helge Kragh, *Cosmology and Controversy,* Princeton University Press, 1996, page 372.

[Questions and conversations with his father on the search for explanations of physical laws such as inertia have -] "…encouraged Richard Feynman, in later years, to question everything, to search for underlying truths, and never to believe that just because some process had been labeled meant that it was understood." See: John and Mary Gribbin, *Richard Feynman: A Life in Science,* A Dutton Book, 1998, page 3.

"The third aspect of my subject is that of science as a method of finding things out. This method is based on the principle that observation is the judge of whether something is so or not. All other aspects and characteristics of science can be understood directly when we understand that observation is the ultimate and final judge of the truth of an idea." Richard P. Feynman, *The Meaning of It All: Thoughts of a Citizen Scientist,* Helix Books, Reading, Mass., 1999, page 15.

"Whereas most theoretical physicists rely on careful mathematical calculation to provide a guide and a crutch to take them into unfamiliar territory, Feynman's attitude was almost cavalier. You get the impression that he could read nature like a book and simply report on what he found, without the tedium of complex analysis." Richard P. Feynman, *Six Easy Pieces: essential of Physics Explained by Its Most Brilliant Teacher:* originally prepared for publication by Robert B. Leighton and Matthew Sands. Addison-Wesley Publishing Company, Reading, Mass., 1994, page xii.

Lee Smolin: *Think About Nature*: "So my impression is that when–Let me come back to how I quoted Feynman: When very smart people have been working under certain assumptions for a very long time–and these ideas have been around for a lot longer than the ideas that Feynman was concerned with were–and we're not succeeding in uncovering new phenomena and new explanations, new understanding for phenomena, it's time to reassess the foundations of our thinking." John Brockman, Editor, *The Universe: Leading Scientists Explore the Origin, Mysteries, and Future of the Cosmos*, Harper Perennial, New York, 2014, pages 140-141.

Feynman on Photons
"Once, though, when I came back from MIT, where I'd been for a few years, he [his father] said to me, 'Now that you've become educated about these things there's one question I've always had that I've never understood very well, ... Well, now. Is the photon in the atom ahead of time, so that it can come out? Or is there no photon in the atom to start with?' ... "I couldn't explain it very well. ... "So he was unsuccessful; he sent me through all these universities in order to find out these things, and he never did find out." See: Christopher Sykes, Editor, *No Ordinary Genius*, W. W. Norton & Company, 1994, page 39.

A.3 – Scientific Theories [explanations]; Standard Models [Particle & Cosmology]; Quantum Field; & General Relativity

"Prediction … is simply no substitute for explanation"

"For even in purely practical applications, the explanatory power of a theory is paramount and its predictive power only supplementary. … Prediction — even perfect, universal prediction — is simply no substitute for explanation." David Deutsch, *The Fabric of Reality: The Science of Parallel Universes — and Its Implications*, Allen Lane, The Penguin Press, New York, 1997, pages 4 & 5.

"A scientific theory is usually felt to be better than its predecessors not only in the sense that it is a better instrument for discovering and solving puzzles but also because it is a better representation of what nature is really like." Thomas S. Kuhn, *The Structure of Scientific Revolutions*, Third Edition, University of Chicago Press, Chicago, 1996, pg. 206.

Understanding Science: How Science Really Works: *Science at multiple levels*: Section:"JUST" A THEORY? " … but in science, a theory is a powerful explanation for a broad set of observations. To be accepted by the scientific community, a theory (in the scientific sense of the word) must be strongly supported by many different lines of evidence." http://undsci .berkeley. edu/article/0_0_0/howscienceworks_19

Standard Models

"The hypothesis of the Big Bang accords with a model in which time and the universe originate jointly. This is the standard model that physics has of the beginning of the universe." Henning Genz, *Nothingness: The science of empty space.* Cambridge, Massachusetts: Perseus books, 1994, page 37.

The New York Review of Books: November 7, 2013 Volume 60, Number 17: by Steven Weinberg Physics: *What We Do and Don't Know*: "In the past fifty years two large branches of physical science have each made a historic transition. I recall both cosmology and elementary particle physics in the early 1960s as cacophonies of competing conjectures. By now in each case we have a widely accepted theory, known as a 'standard model.' " <u>http://www.nybooks.com/articles/ archives/2013/nov/07/physics-what-we-do-and-dont-know</u>

Search: <u>The Physics Hypertextbook: *Beyond the Standard Mode:* Steven Weinberg, 2003</u> – "But it's [standard model] not an entirely satisfactory theory, because it has a number of arbitrary elements. For example, there are a lot of numbers in this standard model that appear in the equations, and they just have to be put in to make the theory fit the observation." <u>http ://physics.info/beyond/</u> or search above title underlined.

"We have pretty good confidence in the ability of the Standard Model to trace the present expansion of the universe back to about a billionth of a second after its supposed start. But when we try to understand what happened earlier than that, we run into the limitations of the model," Steven Weinberg, *Lake Views: This World and the Universe*, The Belknap Press of Harvard University Press, Cambridge, 2009, page 4.

Cornell University Library: P. Pralavorio: *Particle Physics and Cosmology*: "Today, both particle physics and cosmology are described by few parameter Standard Models, i.e. it is possible to deduce consequence of particle physics in cosmology and vice verse." <u>http://arxiv.org/abs/1311.1769</u>

Glenn Elert, The Physics Hypertextbook: T*he Standard Mode:* "The standard model is the name given in the 1970s to a theory of fundamental particles and how they interact. It incorporated all that was known about subatomic particles at the time and predicted the existence of additional particles as well." http://physics.info/standard/

"The development of the ΛCDM (lambda: cold dark matter) model of the universe, also known as the standard model of the big bang cosmology, is a triumph of modern physics. It will be no surprise to learn that this is a development whose origins can be traced back to Einstein." Jim Baggott, *Farewell To Reality: How Modern Physics Has Betrayed the Search for Scientific Truth*, Pegasus Books, New York, 2013, page 104.

Scientific Program: *Cosmology and Particle Physics Beyond the Standard Models* July 30 to August 11, 2007:
"The Standard Model of particle physics and General relativity offer a theoretical framework explaining properties of matter and interactions from subnuclear scales to galactic sizes. However, observational data such as well-controlled measurements of the distribution of large scale structures and cosmic microwave anisotropies indicate that these theories need to be completed at larger energies and larger scales to resolve a number of puzzles: acceleration of the expansion of the universe, nature of the dark matter component of the Universe, origin of the matter-antimatter asymmetry, origin of the inflationary stage in the early Universe." http://servant web.**cern**.ch/servant/cargese/

CERN: *The Standard Model: The Standard Model explains how the basic building blocks of matter interact, governed by four fundamental forces:* "The theories and discoveries of thousands of physicists since the 1930s have resulted in a remarkable insight into the fundamental structure of matter: everything in the universe is found to be made from a few basic building blocks called fundamental particles, governed by four fundamental forces. Our best understanding of how these particles and three of the forces are related to each other is encapsulated in the Standard Model of particle physics." [At CERN, the European Organization for Nuclear Research, physicists and engineers are probing the fundamental structure of the universe.] http://home.web.**cern**.ch/about/physics/stan dard-model

[Planck is a European Space Agency (ESA) space-based observatory] *Planck and The cosmic microwave background:* "The standard model of cosmology rests on the assumption that, on very large scales, the Universe is homogeneous and isotropic, meaning that its properties are very similar at every point and that there are no preferential directions in space." http://www.**ESA**.int/Our_Activities/Space_Science/Planck/Pla nck_and_the_cosmic_microwave_background

Fermilab Inquiring Minds, *The Standard Model of Elementary Particles and Force*, Beyond the Standard Model: "Nevertheless, scientists do not believe that the Standard Model provides complete answers to all our questions about matter." http://www.fnal.gov/pub/science/inquiring/physics/discoveries /top_quark_background/top95_standard_model.html

[At the University of Florida [**ufl**], particle physics and astrophysics joined forces in the Institute for High Energy Physics and Astrophysics, which is centered in the Physics Department.] *High Energy Physics:* "Modern Particle Physics is searching for answers to some of the most fundamental questions in physics, including: What are the truly fundamental constituents of the matter? Are there new particles which would explain the "dark matter" in the Universe? What is the origin of the mass? Are there more than four dimensions in the physics world? ... "Not surprisingly, Particle Physics as a field is becoming inseparable from Astrophysics." http://www.phys.**ufl**.edu /research/hep.shtml

American Institute of Physics, *The Expanding Universe:* "There is more to the advance of science than new observations and new theories. Ultimately, people must be persuaded. In science, as in a court of law, advocates for each side of an issue present the best case possible in an attempt to reach the truth." Copyright ©2015.Brought to you by the Center for History of Physics, a Division of the American Institute of Physics   http://www.aip.org/history/cosmology/ideas/expanding.htm

> That's what this book is all about –to present the best possible case for how the Big Bang Banged.

Quantum Field

What is Quantum Physics?   "That's an easy one: it's the science of things so small that the quantum nature of reality has an effect. Quantum means 'discrete amount' or 'portion'. Max Planck discovered in 1900 that you couldn't get smaller than a certain minimum amount of anything." Higgo, James. *A Lazy Layman's Guide to Quantum* Physics, available at: www.higgo.com/quantum/laymans.htm

University of Nebraska Lincoln (UNL): "QUANTUM THEORY, also quantum mechanics, in physics, a theory based on using the concept of the quantum unit to describe the dynamic properties of subatomic particles and the interactions of matter and radiation."   http://dwb.**unl**.edu/Teacher/NSF/C04/C04Links/www.fwkc.com/encyclopedia/low/articles/q/q021000030f.html

"The word quantum derives from quantity and refers to a small packet of action or process, the smallest unit of either that can be associated with a single event in the microscopic world." http://abyss.uoregon.edu/~js/cosmo/lectures/lec08.html

"Quantum theory is bizarre. In order to try and understand it we need to forget everything we know about cause and effect, reality, certainty, and much else besides. This is a different world, it has its own rules, rules of probability that make no sense in our everyday world. Richard Feynman, the greatest physicist of his generation, said of quantum theory — *'It is impossible, absolutely impossible to explain it in any classical way.'* " [Keith Mayes, *Science, the Universe and God: The Search for Truth*] www.thekeyboard.org.uk/Quantum%20mec hanics.htm

General Theory of Relativity – felt as gravity

Space.Com: Nola Taylor Redd: *Einstein's Theory of General Relativity:* "Einstein then spent ten years trying to include acceleration in the [his Special Relativity] theory and published his theory of general relativity in 1915. In it, he determined that massive objects cause a distortion in space-time, which is felt as gravity." http://www.space.com/17661-theory-general-relativity.html

"The general theory of relativity derives its origin from the need to extend the new space and time concepts of the special theory of relativity from the domain of electric and magnetic phenomena to all of physics and, particularly, to the theory of gravitation." http://abyss.uoregon.edu/~js/21st_century _science/lectures/lec07.html

A. 4 Scientific reporting on the Incredible, the Sensational, the Bizarre, Cognitive Dissonance & Conservation of Ignorance

[See also – Appendix 28 on Superstrings, Hidden Dimensions, and the Multiverse]

the Incredible, the Sensational, the Bizarre
[Sir Martin] "Ryle wrote: 'Cosmologists have always lived in a happy state of being able to postulate theories which had no chance of being disproved' " From: Helge Kragh, *Cosmology and Controversy,* Princeton University Press, 1996, page 309.

"The strangeness of the universe obliges astrophysics to heed the White Queen's words to Alice that she practiced believing six impossible things before breakfast." See: John Noble Wilford, Editor, *Cosmic Dispatches,* W. W. Norton, 2001, page 171.

"Contemporary cosmologists feel free to say anything that pops into their heads. Unhappy examples are everywhere: absurd schemes to model time on the basis of the complex numbers, as in Stephen Hawking's *A Brief History of Time*; bizarre and ugly contraptions for cosmic inflation; universes multiplying beyond the reach of observation; white holes, black holes, worm holes, and naked singularities; theories of every stripe and variety, all of them uncorrected by any criticism beyond the trivial. See: David Berlinski, *Was There a Big Bang?* Commentary Magazine, February 1998, pg. 34-5.

" 'Many people want to believe the incredible, regardless of the evidence,' so noted by Michael Rowan-Robinson; cosmologist at the Imperial College, London - Discussing the French philosopher Montaigne medallion 1576 *Que sais-je?* (what do I know) worn to remind himself that nothing should be believed without evidence." Michael Rowan-Robinson, *The*

Nine Numbers of the Cosmos, Oxford University Press, New York, 1990, page vii.

" … and that when Einstein stopped creating it was because 'he stopped thinking in concrete physical images and became a manipulator of equations.' " See: James Gleick, *Genius*, Pantheon Books, 1992, page 244.

Washington Post: 3/19/2014 by Chris Cillizza: *Americans read headlines. And not much else:* "So, roughly six in 10 people acknowledge that they have done nothing more than read news headlines in the past week. And, in truth, that number is almost certainly higher than that, since plenty of people won't want to admit to just being headline-gazers but, in fact, are." … "The more complex an issue, the less likely it is to break through with a public that really consumes news via headlines and not much else." http://www.washington post.com/blogs/the-fix/wp/2014/03/19/americans-read-headlines-and-not-much-else/

The Incubator: Hatching conversations about science, *Sensationalism in Science, Part I* Posted by Gabrielle Rabinowitz on 1/29/13 "The world according to popular science headlines is a pretty crazy place. Miraculous cures and surprising causes for all of our ailments can be found in a bite of chocolate. Everything we knew about something is wrong. And NASA has confirmed either the existence of aliens or a world-engulfing black hole." http://incubator.rockefeller .edu/?p=172

David F. Ransohoff, MD: *Sensationalism in the Media: When Scientists and Journalists May Be Complicit Collaborators:* "While different styles of communication may contribute to inaccurate science journalism, we believe that subtle incentives sometimes cause scientists, journalists, and others involved in the reporting of science to contribute to sensationalism. ... when a finding is reported in a sensational way, the results may create a national media feeding frenzy." http://ecp.acponline.org/julaug01/ranso hoff.pdf

"You have 30 seconds to tell me why I should care about what you have to say." ... "Headlines/titles: Tabloids take the cake for grabbing your attention with headlines. They're outlandish, exaggerated, and preposterous, but we remember them." https://www.servicescape.com/article.asp?cid=93303

"Episodic reporting [containing or consisting of a series of loosely connected parts or events] sensationalizes information to seize reader's attention, while neglecting significant details that might provide a context for understanding events." See Sharon L. Nichols, Thomas L. Good, *America's Teenagers-- Myths and Realities*: *Media Images, Schooling, and the Social Costs of Careless Indifference,* Routledge, Mahwah, N.J., 2004, page 56.

Attracting Readers vs. Sensationalizing Crime, Robin L. Barton, April 4, 2011: "The word "sensationalism" was first used in the nineteenth century to criticize the so-called "yellow journalism" of newspapers such as the *New York World* and *New York Journal*. The term has since been applied to media coverage that's controversial, shocking, attention-grabbing, graphic, appealing to the lowest instincts or focused on superficial details." http://www.thecrimereport.org/view points/robin-barton/2011-04-attracting-readers-vs-sensationali zing-crime

"Complex subjects and affairs are often subject to sensationalism. Exciting and emotionally charged aspects can be drawn out without providing the elements needed (such as pertinent background, investigative, or contextual information) for the audience to form its own opinions on the subject." http://en.wikipedia.org/wiki/Sensationalism

"Sensationalism is content that is: • Controversial • Shocking • Attention grabbing • Failing to explain the broader issues behind the story while focusing on superficial details • Published to attract readers, regardless of whether the information is accurate or informative …" http://freelance-writing.lovetoknow.com/Journalism_and_Sensationalism

"Most of us assume we are seeing the world the way it really is." http://www.az_quotes.com/author/40598-Ned_Herrmann "Ned Herrmann, His first widely acclaimed book, 'The Creative Brain,' traced the scientific and historical roots of his innovative Whole Brain® Thinking approach." http://www.herrmannsolutions.com/about/

Cognitive Dissonance
The Huffington Post *Why Congress' Obamacare Doomsday Cult Can't Admit It Was Wrong,* Posted: 03/16/2015 7:32 am EDT, by Michael McAuliff **– at about the 24th paragraph:** "The way cognitive dissonance works is that when people are confronted with information that contradicts either their beliefs or actions, they feel discomfort. To feel better, they either have to modify their beliefs and actions, or find some way to discount the disconfirming information. And the more effort someone invests in a particular action or idea, the greater the lengths they will go in crafting justifications to ease their discomfort. "That's really the way the human mind works, even outside of Washington," [Elliot] Aronson told HuffPost".
"Indeed, committing to a specific ideology can make it much harder to see facts clearly, let alone acknowledge them. [Elliot] Aronson noted that it's especially hard for people who spent the last five years opposing a specific policy." Michael McAuliff, http://www.huffingtonpost.com/2015/03/16/obamacare-cognitive-dissonance_n_6866278.html

"It turns out that the scientist behaves the way the rest of us do when our beliefs are in conflict with the evidence. We become irritated, we pretend the conflict does not exist, or we paper it over with meaningless phrases." See: Robert Jastrow, *God and the Astronomers.* New York: Warner Books, 1978, page 5.

The Conservation of Ignorance

"The Mathematician Georg Cantor spoke of a law of conservation of ignorance. A false conclusion once arrived at and widely accepted is not easily dislodged and the less it is understood the more tenaciously it is held." See: Morris Kline, *Mathematics: The Loss of Certainty.* Oxford University Press, 1980, page 88.

"This is another general fact about scientific explanation: if one has a misconception, observations that conflict with one's expectations may (or may not) spur one into making further conjectures, but no amount of observing will **correct** the misconception until after one has thought of a better idea; in contrast, if one has the right idea one can explain the phenomenon even if there are large errors in the data." See: David Deutsch, *The Beginning of Infinity,* Viking, 2011, page 18.

"He [Thomas Kuhn] said that scientists worked within the confines of their theories long after there was enough factual proof to disprove them. A new theory was embraced only when someone finally overturned the whole shebang completely,..." See: Karen C. Fox, *The Big Bang Theory: What it is, Where it Came From, and Why it works.* John Wiley & Sons, Inc. 2002, page 11.

A. 5 Cosmological Assumptions Currently in Force – created during the years 1915-34.

Regarding the elegant symbolic mathematics of general relativity, we find that "After he published his famous paper in 1916, Einstein later conceded that the mathematical difficulties of his General Theory of Relativity were a 'very serious' impediment to its further development." http://arc hive.ncsa.illinois.edu/Cyberia/NumRel/EinsteinEquations.html

Attempts to apply GR to our Universe as a whole required an assumption of homogeneity or uniform distribution of how matter existed in our Universe. Even with such an assumption, "Einstein from before 1920 until his death in 1955, struggled to find laws of physics far more general than any known before" without success. http://www.aip.org/his tory/exhibits/einstein/philos1.htm

This is evidenced by the work published by others starting with Willem de Sitter's mathematical solutions to Einstein's field equations in the absence of matter, maintaining that relativity actually implied that the space was expanding with no matter, no planets, no galaxies. Hence we have —

space expanding today with no supporting physics:
"In a letter to [Willem de] Sitter—discovered in a box of old records in Leiden a few years ago—Einstein wrote, 'This circumstance (of an expanding Universe) irritates me,' and in another letter about the expanding Universe [de Sitter's math in 1917 described expanding space, but without any matter in such expanding space], 'To admit such possibilities seems senseless.'" See: Robert Jastrow, *God and the Astronomers.* New York: Warner Books, 1978, page 17.

Here is where our **'creation from nothing'** comes from, the next assumption – mythology married to mathematical concepts – along with a desire to support this math with the fact of our Universe's existence – **Hindu mythology:**

From Alexander A Friedmann, who published his 1923 rework of Einstein's relativity, we note that in his book, Friedmann formulates the results of his first cosmological paper in just a few sentences: ... "This brings to mind what the Hindu mythology has to say about cycles of existence, and it also becomes possible to speak about 'the creation of the world from nothing." This effectively denied any existence of a God. Reported in the 1993 book: *"Alexander A Friedmann: The Man Who Made the Universe Expand"* page 156-7 by Tropp, Frenkel, and Chernin (Cambridge University Press, Cambridge.

To complete the mathematics, spawned off of relativity, this Abbé Georges-Henri Lemaître published that 1931 notation that spawned our concept of a singularity with **Ifs** about creation from a single quantum; from reworked math and with no physical evidence —— just **Ifs.** In his letter to the journal *Nature* in 1931, Lemaître noted that "If the world has begun with a single quantum, the notions of space and time would altogether fail to have any meaning at the beginning…if this suggestion is correct, the beginning of the world happened a little before the beginning of space and time." See *The Beginning of the World from the point of View of Quantum Theory.* Nature 127,706, 1931.   [**if** is not science]   http://ww w.nature.com/nature/journal/v127/n3210/full/127706b0.html

"with the help of assumptions"

"Sir James Jeans' question— … "Is the universe expanding at about the rate indicated by the 'spectra of the nebulae?' To Lemaître there was no doubt. He considered the cosmic expansion a scientific fact for which an oracle was not needed: 'The expansion of the universe is a matter of astronomical facts interpreted by the theory of relativity, with the help of assumptions as to the homogeneity of space, without which any theory seems to be impossible.' " See: Helge Kragh, *Cosmology and Controversy,* Princeton University Press, 1996, page 49.

That **singularity out of nothing** is continually repeated like a mantra, repeated even by NASA. See: *Foundations of Big Bang Cosmology* "…it was born with zero volume and grew from that" [That singular nothing in this concept means not even empty space and without space this nothing is the complete absence of anything, a very difficult concept to imagine.] http://map.gsfc.nasa.gov/universe/bb_concepts.htm

The **next assumption — Cosmological Principle** [A. 21] stating that Earth has no special place in our Universe. Milne's homogeneous Universe has no center and no edge according to his mathematics.

In 1932-4, Edward A. Milne formulated the "*cosmological principle*" describing a homogenous universe that looks the same regardless of one's position in our Universe. "According to Milne's principle, every observer in the universe should get the same world picture, that is, should make precisely the same observations of the universe at the same moment as any other observer.(Milne 1934b)" at: http://plato.stanford.edu/entries/cosmology-30s/#ButMilEnd

"Einstein's equations are difficult to solve, and the only way to make progress is to put in some simplifying assumptions. This is the first opportunity, within modern cosmology, for theoretical prejudices to make an appearance. The standard assumption or prejudice, is that material is, on the average, scattered uniformly throughout space. Obviously, this is wrong." See: David Lindley, *The End of Physics*, Basic Books, 1993, page 137.

A. 6 – The Singleton [Singularity] [Dot] [•]

"what existed before the Big Bang, has scientists baffled"
John P. Millis, Ph.D: *The origin of the Universe: The Big Bang:* " ... what existed before the Big Bang, has scientists baffled. By definition, nothing existed prior to the beginning, but that fact creates more questions than answers. For instance, if nothing existed prior to the Big Bang, what caused the singularity to be created in the first place?"   http://space .about.com/od/astronomybasics/a/Origin-Of-The-Universe.htm

"Singularities... defy our current understanding of physics"
"According to the standard theory, our universe sprang into existence as [a] "singularity" around 13.7 billion years ago. What is a "singularity" and where does it come from? Well, to be honest, we don't know for sure. Singularities are zones which defy our current understanding of physics." From All About Science.org. http://www.big-bang-theory.com/#sthash w0SqUcRX.dpuf

"It is difficult enough to imagine a time, roughly 13.7 billion years ago, when the entire universe existed as a singularity. According to the big bang theory, one of the main contenders vying to explain how the universe came to be, all the matter in the cosmos -- all of space itself -- existed in a form smaller than a subatomic particle."   http://science.howstuffworks. com/dictionary/astronomy-terms/before-big-bang.htm

"At a point in time, about 13.7 billion years ago all matter was compacted into a very small ball with infinite density, and intense heat called a singularity." http://www.universetoday .com/54756/what-is-the-big-bang-theory/

" ... *how* did it come into existence, and *what* existed before it?"... "Most scientists now believe that the answer to the first part of the question is that the Universe sprang into existence from a singularity -- a term physicists use to describe regions of space that defy the laws of physics." http://space. about.com/od/astronomybasics/a/Origin-Of-The-Universe.htm

"our knowledge here is very shaky"

"It is often said – in both popular-level books and in textbooks – that this singularity marks the beginning of time itself. Perhaps it's so, but any honest cosmologist would admit that our knowledge here is very shaky. The extrapolation to arbitrarily high temperatures takes us far beyond the physics that we understand, so there is no good reason to trust it. The true history of the universe, going back to 't = 0', remains a mystery that we are probably still far from unraveling." See: Alan H.Guth, *The Inflationary Universe,* Basic Books, 1998, pages 86-7.

"The closer we get to the Big Bang, going backward in time, the hotter all matter must have been. At the instant of the Big Bang, the universe was infinitely hot and infinitely dense." See: Henning Genz, *Nothingness*, Perseus books, 1994, page 38.

A. 7 – Universe out of Nothing

"Science tells us nothing about – our universe's earliest instant"
"Science tells us nothing about the way space, time and matter behaved in our universe's earliest instant, from the time of the Big Bang to 10 $^{-43}$ seconds later." http://archive.ncsa. illinois.edu/Cyberia/Cosmos/InTheBeginning.html

"Big Bang Theory - The Premise: "Discoveries in astronomy and physics have shown beyond a reasonable doubt that our universe did in fact have a beginning. Prior to that moment there was nothing; during and after that moment there was something: our universe." http://www.big-bang-theory.com/#s thash.TV0cpjNC.dpuf

"nothing ... somehow turned into something"
How the universe appeared from nothing, July 2011, "Many physicists now believe that the universe arose out of nothingness during the Big Bang which means that nothing must have somehow turned into something." http://www. newscientist.com/blogs/nstv/2011/07/how-the-universe-appea red-from-nothing.html

What came before the big bang? "There was no such epoch as 'before the big bang,' because time began with the big bang, says physicist and astrobiologist Paul Davies " *Nothing: Surprising Insights Everywhere from Zero to Oblivion, published by The Experiment,* NY, March 25, 2014 Jeremy Webb (Editor) http://boingboing.net/2014/05/20/what-came-before-the-big-bang.html

"The singularity didn't appear *in* space; rather, space began inside of the singularity. Prior to the singularity, *nothing* existed, not space, time, matter, or energy - nothing. So where and in what did the singularity appear if not in space? We don't know. We don't know where it came from, why it's here, or even where it is. http://www.big-bang-theory.com/#sthash .TV0cpjNC.dpuf

what we know is still only speculation
"How did our universe begin? How old is our universe? How did matter come to exist? …much time and effort has been spent looking for some clue. Yet…, much of what we know is still only speculation." www.umich.edu/~gs265/bigbang.htm

"How did the universe begin? … "The big bang, however, is merely a global description of the origin of the universe. "The goal of physics today is … understand what exactly happened around the moment of the big bang to get the universe started." http://www.ugcs.caltech.edu/~yukimoon/BigBang/

"The Big Bang theory is very useful in explaining the events that took place after the birth of the universe, but scientists began to search for answers for events before then." http://w ww.oglethorpe.edu/faculty/~m_rulison/Astronomy/Group/Fall %2099/expansion_of_the_universe.htm

Big Bang Didn't Need God, Stephen Hawking Says, Rod Pyle, Space.com, Contributor: "M-theory posits that multiple universes are created out of nothing," http://www.space.com /20710-stephen-hawking-god-big-bang.html

"How did it happen that at some point in time, something appeared out of nothing? To this day, physics cannot give a definite answer to this question." Henning Genz, *nothingness: the science of empty space*, Perseus Publishing, Cambridge, Massachusetts, 1998, page viii.

A Universe from Nothing? By Jake Hebert, PhD: "Theoretical physicist Lawrence Krauss presented in a recent book his claim that the laws of physics could have created the universe from nothing. Likewise, other physicists offer similar arguments." http://www.icr.org/article/universe-from-nothing

"Krauss: ... I don't think I argued that physics has definitively shown how something could come from nothing; physics has shown how plausible physical mechanisms might cause this to happen." http://www.theatlantic.com/technology/archive/2012/04/has-physics-made-philosophy-and-religion-obsolete/256203/

A. 8 – The Big Bang's Creation of Time, Space & Matter

No arguments supporting the Big Bang –
"The key message we hope to have transmitted to the reader in this book is that there are neither observational data nor incontrovertible theoretical arguments to support the belief that the Big Bang represents the beginning of the Universe." See Maurizio Gasperini, *The Universe Before the Big Bang: Cosmology and String Theory*, Springer, Berlin, 2008, page 193.

"What existed prior to this event is completely unknown"
"About 15 billion years ago a tremendous explosion started the expansion of the universe. This explosion is known as the Big Bang. At the point of this event all of the matter and energy of space was contained at one point. What existed prior to this event is completely unknown and is a matter of pure speculation."   http://www.umich.edu/~gs265/bigbang.htm

"According to most astrophysicists, all the matter found in the universe today -- including the matter in people, plants, animals, the earth, stars, and galaxies -- was created at the very first moment of time, thought to be about 13 billion years ago."   exploratorium.edu/origins/cern/ideas/bang.html

"no observational consequences"
Stephen Hawking: *The Beginning of Time:*   "Since events before the Big Bang have no observational consequences, one may as well cut them out of the theory, and say that time began at the Big Bang." http://www.hawking.org.uk/the-begin ning-of-time.html

"Most cosmologists believe that the Universe, and with it space and time, exploded into being some 13.7 billion years ago at the Big Bang, and that it has been expanding ever since." http://www.nature.com/news/2010/101210/full/news.2010.665.html

"Time, space and matter all began with the Big Bang." http ://www.esa.int/esaKIDSen/SEMSZ5WJD1E_OurUniverse_0.html

"The big bang is not like an explosion of matter in otherwise empty space; rather, space itself began with the big bang and carried matter with it as it expanded. Physicists think that even time began with the big bang. Today, just about every scientist believes in the big bang model." http://www.ugcs.caltech.edu/~yukimoon/BigBang/

exploded with unimaginable force
"The universe began, scientists believe, with every speck of its energy jammed into a very tiny point. This extremely dense point exploded with unimaginable force, creating matter and propelling it outward to make the billions of galaxies of our vast universe. Astrophysicists dubbed this titanic explosion the Big Bang." http://www.exploratorium.edu/origins/cern/ideas/bang.html

"According to the big bang theory, the universe began by expanding from an infinitesimal volume with extremely high density and temperature. The universe was initially significantly smaller than even a pore on your skin." http://www.ugcs.caltech.edu/~yukimoon/BigBang/

A. 9– Mathematics is not Physics.   Postulates, Premise, Axiom, Equations, & an extra Six Dimensions

"Not all of science is mathematical.   Not everything mathematical is science."   See: David Lindley, *The End of Physics*, Basic Books, NY, 1993, page 13.

"Physics is not mathematics, and mathematics is not physics. One helps the other.   But in physics you have to have an understanding of the connections of words with the real world." Richard P. Feynman, *The Character of Physical Law*, M.I.T. Press, Cambridge, Massachusetts, 1965, page 55.

That was just mathematics,   …   not reality
"Einstein published a correction to *his* correction of Friedmann's paper, … That was just mathematics, according to Einstein, not reality."   Pedro G. Ferreira, *The Perfect Theory: A Century of Geniuses and the Battle over General Relativity*, Houghton Mifflin Harcourt, Boston, 2014, p 34, 35.

"Mathematics and physics are seen as two sides of the same coin. They are quite different, however. The truth of physics and other natural sciences can only be established through observation and experiment." From Einstein's *Geometry and Experience* as reported in his book – Stephen Hawking, *A Stubbornly Persistent Illusion*, Running Press, Philadelphia, 2007, page 247.

"In my opinion the answer to this question is, briefly, this: As far as the laws of mathematics refer to reality, they are not certain; and as far as they are certain, they do not refer to reality."   Albert Einstein; G B Jeffery; Wilfrid Perrett, *Sidelights on relativity. I. Ether and relativity. II. Geometry and experience*, Methuen, London, 1922,   also available at: http://archive.org/stream/sidelightsonrela00einsuoft/sidelights onrela00einsuoft_djvu.txt

"Physics is really a different *game* from mathematics. It's not just that it has laxer rules then the game of mathematics, or that it is an *easier version* of mathematics. Rather, it has *different* rules altogether, so someone who was trained to play the game of mathematics would have difficulty transferring what they've acquired in practicing mathematics to a different game." http://aeolist.wordpress.com/2007/03/17/physics-as-applied-mathematics-not/

"Mathematics deals exclusively with the relations of concepts to each other without consideration of their relation to experience. Physics too deals with mathematical concepts; however, these concepts attain physical content only by the clear determination of their relation to the objects of experience." From Einstein's *Out of my Later Years,* as reported by Stephen Hawking in his book – *A Stubbornly Persistent Illusion*, Running Press, 2007, page 385.

"Even within the community of particle physicists there are those who think that the trend toward increasing abstraction is turning theoretical physics into recreational mathematics, endlessly amusing to those who can master the techniques and join the game, but ultimately meaningless because the objects of the mathematical manipulations are forever beyond the access of experiment and measurement." See: David Lindley, *The End of Physics*, Basic Books, 1993, page 19.

"He [Steven Weinberg] was careful to acknowledge that neither the superconducting supercollider nor any other earthly accelerator could provide direct confirmation on a final theory; physicists would eventually have to rely on mathematical elegance and consistency as guides." See: John Hogan, *The End Of Science*: *Facing The Limits Of Knowledge In The Twilight Of The Scientific Age,* Helix Books, Reading, Mass., 1996, page 73.

"We have an assumption now that's getting stronger and stronger that mathematics is the only way to deal with reality," Bohm said. [David Bohm is widely considered to be one of the most significant theoretical physicists of the 20th century.] See: : John Hogan, *The End Of Science*: *Facing The Limits Of Knowledge In The Twilight Of The Scientific Age,* Helix Books, Reading, Mass., 1996, page 88.

"Finally, it seems clear that we will never be able to explain our most fundamental scientific principles.   ...   The only kind of explanation I can imagine (if we are not just going to find a deeper set of laws, which would then just push the question farther back) would be to show that mathematical consistency requires these laws. But this is clearly impossible, because we can already imagine sets of laws of nature that, as far as we can tell, are completely consistent mathematically but that do not describe nature as we observe it."   Steven Weinberg, *Lake Views: This World and the Universe*, The Belknap Press of Harvard University Press, Cambridge, 2009, page 22.

"Said to Lemaître by Einstein, [in 1927] 'Your calculations are correct, but your physics is abominable.' And later retracted."   See: Martin Gorst, *Measuring Eternity: The Search for the Beginning of Time,* Broadway book, New York, 2001, pages 233 &236.

After Lemaître in 1933 detailed his Big Bang theory, Einstein stood up applauded, and said, "This is the most beautiful and satisfactory explanation of creation to which I have ever listened." J. P. McEvoy, *A Brief History of the Universe*, Robinson Publishing, London, 2010, page 246.

"Big Bang cosmology [referring to 20th century's cosmology] does not begin with observations but with mathematical derivations from unquestionable assumptions."   See: Eric J.

Lerner, *The Big Bang Never Happened: a Startling Refutation of the Dominant Theory of the Origin of the Universe.* Time Books, 1990, page 114.

Science & Math Assumptions that sound like fact:

Postulates
Inaccurate predictions derived from flawed **postulates**
"In reading a postulate, do two things. First, try to understand and appreciate the basic physical idea embodied in the postulate; this idea will ultimately be important in understanding the macroscopic properties ... Second, identify possible weakness or flaws in the postulates. Inaccurate predictions by a theory derive from flawed postulates used in the derivation of the theory." www.chm.davidson.edu/vce /kineticmoleculartheory/basicconcepts.html

[In Mathematics, Logic]: "–a proposition that requires no proof, being self-evident, or that is for a specific purpose assumed true, and that is used in the proof of other propositions; axiom." http://dictionary.reference.com/browse /postulate

Premise
[In Foundations of Mathematics; Logic; General Logic;] "A premise is a statement that is assumed to be true. Formal logic uses a set of premises and syllogisms to arrive at a conclusion." http://mathworld.wolfram.com/Premise.html

Axiom
[in Logic, Mathematics.]   "a proposition that is assumed without proof for the sake of studying the consequences that follow from it." http://dictionary.reference.com/browse/axiom

"*Axioms* are *not* self-evident truths in any sort of rational system, they are *unprovable assumptions* whose truth or falsehood should always be mentally prefaced with an implicit 'If we assume that...' "   http://www.phy.duke.edu/~rgb/Phil osophy/axioms/axioms/node27.html

"An axiom is a proposition regarded as self-evidently true without proof."   http://mathworld.wolfram.com/Axiom.html

Equations are only as good as the assumptions

"…   equations are only as good as the assumptions you make in formulating them. If the assumptions are false and do not reflect nature, the equations will lead to wrong conclusions even if they are set forth by the greatest scientific mind in the world." Amir D. Aczel, *God's Equation: Einstein, Relativity, and the Expanding Universe*, Four Walls Eight Windows, New York, 1999, page 101.

an extra six dimensions

"Though it is a strange concept physically, it is the predictive power of equations that interest scientists, rather than their comprehensibility – and an extra six dimensions do not constitute an insurmountable problem, mathematically." See – Seife, Charles, *Zero: the biography of a dangerous idea,* Viking, New York, 2000, page 197.

A. 10 –Jargon

"I see jargon as one of the biggest barriers in the way. Many blogs fall into the trappings of scientific writing: passive voice, labored constructions, and roundabout sentences … And then, there are the words themselves. … Writers should always remember that the more technical you get, the more restrictive you get, even if people are writing for a scientific audience." http://blogs.discovermagazine.com/notrocket science/2010/11/24/on-jargon-and-why-it-matters-in-science-writing/#.VB8ovWMVP_d

"Even between scientists, you hear criticism about the amount of jargon in talks and papers. I have heard several times that community ecology is a frequent offender when it comes to over-reliance on jargon (definition: "words or expressions that are used by a particular profession or group and are difficult for others to understand")." http://evol-eco.blogspot. com/2013/06/speaking-language-is-jargon-always-bad.html

"Many words that seem perfectly normal to scientists are incomprehensible jargon to the wider world. And there are usually simpler substitutes." http://freakonomics.com/2011/ 10/25/jargon-fail-why-scientists-are-bad-at-crafting-simple-clear-messages/

"…the power of some words is actually disguising the true nature of an idea or intent. Sometimes this is intentional, and sometimes unintentional, but either way the result is people being mislead or alienated by language. … One type of language that can be intrinsically misleading or (more often) simply incomprehensible is Jargon. Jargon is defined as vocabulary which is meaningful only to a particular profession or group. …. The problem arises when the group that understands the language either forgets that they are basically speaking gibberish to everyone else, or worse, when they use jargon terms in the full knowledge that they are basically gibberish to everyone else." http://radiofreethinker.com /2011/07/18/come-again-jargon-in-scientific-communication

"There's no reason for the jargon when you're trying to communicate the essence of the science to the public [and professionals, see next quote about James] because you're talking what amounts to gibberish to them," [Alan] Alda said in a recent interview with The Associated Press." http://phys .org/news /2013-05-alan-alda-scientists-jargon.html #jCp

"Over the years, I [James] have attended many physics colloquia that were too difficult to follow — I've rarely heard one that was too simple. It is a sad truth, but expertise in physics research does not automatically make one an expert at communicating physics. Too often one leaves a colloquium more turned off than excited following a presentation on a cutting-edge breakthrough. This is a missed opportunity for both the speaker and the audience, but it need not be this way." James Kakalios, the Taylor Distinguished Professor in the School of Physics and Astronomy at the University of Minnesota, *The Physics of Physics Colloquia,* APSNEWS, February 2015, page 8. http://www.aps.org/publications/aps news/201502/backpage.cfm

A very good example of this 'Jargon' difficulty is a quote in this book starting on page 177 – the quote is so technical it is virtually indecipherable.

A. 11 – the limits of science, how the laws of nature are derived, & the scientific method

Dr. David Stewart, *The Limits of Science in a Nutshell,*
"#1 of 19 – Science explains nothing; it can only describe."
http://healthimpactnews.com/2014/the-limitations-of-science-2/

"As a famous scientist once said, 'Smart people (like smart lawyers) can come up with very good explanations for mistaken points of view.' In summary, the scientific method attempts to minimize the influence of bias or prejudice in the experimenter when testing an hypothesis or a theory." http://teacher.nsrl.rochester.edu/phy_labs/appendixe/appendixe.html

" ... in science, a theory is something that is very well supported by observation and experimentation." http://www.livescience.com/20896-science-scientific-method.html

"The scientific method is a process for forming and testing solutions to problems, or theorizing about how or why things work." http://www.studygs.net/scimethod.htm

"Science is based purely around observation and measurement, and the vast majority of research involves some type of practical experimentation. ... This process of induction and generalization allows scientists to make predictions about how they think that something should behave, and design an experiment to test it. ... This experiment does not always mean setting up rows of test tubes in the lab or designing surveys. It can also mean taking measurements and observing the natural world. ... The scientific method uses some type of measurement to analyze results, feeding these findings back into theories of what we know about the world." https://explorable.com/what-is-the-scientific-method

"The history of science has been a process of gradually discovering and revising the laws of nature. Often, we discover new laws by making guesses, ... But then we must test those laws against experiment. ... As we develop new measuring devices, make better experiments, and reconceive our ideas of scientific principles, we constantly update and revise what we hold to be the laws of nature." Alan Lightman, *The Accidental Universe: The World You Thought You Knew*, Pantheon Books, New York, 2013, pages 56 & 57.

How the Scientific Method Works by William Harris
"And yet, used properly, the scientific method is one of the most valuable tools humans have ever created. It helps us solve everyday problems around the house and, at the same time, helps us understand profound questions about the world and universe in which we live." http://science.howstuffworks.com/innovation/scientific-experiments/scientific-method10.htm

"Nature's laws are man's creation. We, not God, are the lawgivers of the universe. A law of nature is man's description and not God's prescription." See: Morris Kline, *Mathematics: The Loss of Certainty.* Oxford University Press, 1980, page 98.

"The natural world is not objectively given to us. It is man's interpretation or construction based on his sensations, and mathematics is a major instrument for organizing the sensations." See: Morris Kline, *Mathematics: The Loss of Certainty.* Oxford University Press, 1980, page 341.

"But science at its deepest level is an intensely creative activity, just like the arts." See: Gregory J. Chaitin, *Conversations with a Mathematician: Math, Art, Science and the limits of Reason.* Great Britain: Springer, 2002, page 52.

"But, in reality, scientific theories are not 'derived' from anything. We do not read them in nature, nor does nature write them to us. They are guesses – bold conjectures. Human minds create them by rearranging, combining, altering and adding to existing ideas with the intention of improving upon them." See: David Deutsch, *The Beginning of Infinity,* Viking, 2011, page 4.

"Scientific explanations [or theories] are about reality, most of which does not consist of anyone's experiences." … "Science often predicts – and brings about – phenomena spectacularly different from anything that has been experienced before." See: David Deutsch, *The Beginning of Infinity,* Viking, 2011, page 6.

" … science is a procedure for the invention and evaluation of hypotheses that may be used to explain why things happen as they do. … scientific explanations are always tentative proposals, offered in hopes of capturing the best outlook on the matter but subject to evaluation, modification, or even overturn in light of further evidence." http://www.philosophy pages.com/lg/e15.htm

"Physics constitutes a logical system of thought which is in a state of evolution, and whose basis cannot be obtained through distillation by any inductive method from experiences lived through, but which can only be attained by free invention. The justification (truth content) of the system rests in the proof of usefulness of the resulting theorems on the basis of sense experiences," From Einstein's *Out of my Later Years,* as reported by Stephen Hawking in his book – *A Stubbornly Persistent Illusion*, Running Press, 2007, page 434.

"Einstein wrote (1946): The first point is obvious: the theory must not contradict empirical [verifiable] facts. However evident this demand may be in the first place appear, its application turns out to be quite delicate." See: Arthur I. Miller, *Albert Einstein's Special Theory of Relativity: Emergence (1905) and Early Interpretation (1905—1911)*. Addison-Wesley Publishing Company, Inc. 1981, page 124.

BASIC LIMITATIONS OF SCIENTIFIC KNOWLEDGE
Adapted from "*Nature of Modern Science and Scientific Knowledge*" by Martin Nickels, Professor of Anthropology.
"A. Our senses have their own biological limitations. ...
"B. Our mental processing of sensory data is not always reliable. ...
"C. It's impossible to know if we have considered all possible alternative explanations.
"D. Scientific knowledge is necessarily contingent knowledge (and therefore uncertain). ... It is dependent on available evidence, circumstances, tools and our analysis.
"NEVERTHELESS, scientific knowledge is the most reliable knowledge we can have about the natural world and how it works." http://www.indiana.edu/~ensiweb/NOS%20 Over.BasicAssump.html

"To tell the truth, we still don't understand what light and matter, with their atoms and particles, really are." See: Harald Fritzsch, *An Equation That Changed the World: Newton, Einstein, and the Theory of Relativity,* The University of Chicago Press, 1994, page 92.

"Science cannot yet really 'explain' electricity, magnetism, and gravitation; their effects can be measured and predicted, but of their ultimate nature no more is known to the modern scientist than to Thales of Miletus, who first speculated on the electrification of amber around 585 B.C." See: Lincoln Barnett, *The Universe and Dr. Einstein: with a forward by Albert Einstein.* Mattituck, American Reprint Company, N.Y., 1950, page 16.

"When we say that one truth explains another, as for instance the physical principles (the rules of quantum mechanics) governing electrons in electric fields explain the law of chemistry, we do not necessarily mean that we can actually deduce the truths we claim have been explained." See Steven Weinberg, *Dreams of a Final Theory*. New York: Pantheon Books, 1992, page 9.

we have no knowledge of what energy is

"It is important to realize that in physics today, we have no knowledge of what energy is." See: Richard Feynman, *Six Easy Pieces,* Helix Books, 1963, page 71.

Observation:

Science finds the tools

Philosophy seeks the Craftsman

A. 12 – The Amazing Atoms & Star Power

·All atoms created during the Big Bang[see 1's next page]
·No atoms break down by themselves
·Therefore all atoms are at least 13.8 billion years old

·Basic parts – electron & nucleus (made of protons & neutrons [see 2's]; these are also called nucleons confined by 'strong atomic force') - no light/photons found in atom

·light/photons propelled at the speed of light – that 186,282 miles/ second during a chemical or atomic reaction [see 3's]

·electrons [see 4's] – negatively charged & repel electrons
·electrons – attract positively charged protons
·undescribed force repels electrons from protons

·protons [see 5's] – positively charged, repel protons and are made of over 400 different kinds of sub atomic particles including quarks, too many of these to count, confined in and traveling inside at near the speed of light, all bound together by something called the strong interaction [see 6's]

·neutrons – no electro-magnetic charge but otherwise has the same sub atomic construction as the protons [see 6's]

·Standard Model [see 7's] describes sub-atomic activity

·the basic properties of a simple proton include – spin [see 8's]

·binding of atoms [see 9's]

The fuel of stars - Fission and Fusion
·Conversion of U235 atoms into energy [see 10]

1 "In the very beginning, both space and time were created in the Big Bang. It happened 13.7 billion years ago. Afterwards, the universe was a expanding soup of fundamental particles." http://www.haystack.mit.edu/edu/pcr/Astrochemistry/3%20-%20MATTER/nuclear%20 synthesis.pdf

[1] "The Big Bang theory predicts that the early universe was a very hot place. One second after the Big Bang, the temperature of the universe was roughly 10 billion degrees and was filled with a sea of neutrons, protons, electrons, anti-electrons (positrons), photons and neutrinos." http://map.gsfc.nasa.gov/universe/bb_tests_ele.html

[1] "You and everything around you, every single natural and man-made thing you can see, every rock, tree, butterfly, and building, comprises atoms that originally arose during the Big Bang." www.pbs.org/wgbh/nova/space/star-in-you.html

[1] "HAVING vainly sought supporting evidence for more than a decade, most physicists have abandoned a once popular theory that the proton, a fundamental building block of atoms, spontaneously disintegrates." http://www.nytimes.com/1992/09/29/science/proton-decay-theory-may-be-in-for-revival-from-new-approach.html

[1] "Super-Kamiokande is a neutrino observatory located under Mount Kamioka, Japan. The observatory was designed to search for proton decay etc. Kamiokande did not achieve its primary goal, the detection of proton decay." http://en.wikipedia.org/wiki/Super-Kamiokande

[2] "Atoms - The basic chemical building block of matter, atoms are composed of electrons, protons, and neutrons." http://physics.about.com/od/atomsparticles/a/particles.htm

122

[2] "All stable *atoms* except hydrogen contain some number of *neutrons.*" http://www.universetoday.com/82128/parts-of-an-atom/

[2] "Look at a periodic table. The atomic number (hydrogen 1, helium 2, etc) is the number of protons. The atomic weight (hydrogen 1, helium 4, etc) is the sum of the protons and neutrons." http://www2.estrellamountain.edu/faculty/farabee/biobk/biobookchem1.html

[2] "The nucleus of an atom is composed of roughly equal numbers of protons and neutrons which are held together by the nuclear force. Protons and neutrons are in turn made out of point-like particles called quarks and gluons. We have a fairly good understanding of how quarks and gluons interact to form protons and neutrons, and the nominal structure of atomic nuclei in terms of protons and neutrons has been accurately understood for many years. But what we do not yet understand is how a system of quarks and gluons in a large volume behave, such as existed in the very early Universe a few micro-seconds after the Big Bang." www.rhip.utexas.edu/

[2] Jared Sagoff, March 1, 2012: *New picture of atomic nucleus emerges:* " When most of us think of an atom, we think of tiny electrons whizzing around a stationary, dense nucleus composed of protons and neutrons, collectively known as nucleons. A collaboration between the U.S. Department of Energy's Argonne and Thomas Jefferson National Laboratories has demonstrated just how different reality is from our simple picture, showing that a quarter of the nucleons in a dense nucleus exceed 25 percent of the speed of light, turning the picture of a static nucleus on its head." www.anl.gov/articles/new-picture-atomic-nucleus-emerges

No evidence of photons existing in atoms by any source.

[3] "Under the photon theory of light, a *photon* is a discrete bundle (or *quantum*) of electromagnetic (or light) energy. Photons are always in motion and, in a vacuum, have a constant speed of light to all observers, at the vacuum speed of light (more commonly just called the speed of light) of c = 2.998 x 10^8 m/s." = [meters/second] or 186,282 miles/sec. http://physics.about.com/od/lightoptics/f/photon.htm

[3] "There are many different ways to produce photons, but all of them use the same mechanism inside an atom to do it. This mechanism involves the energizing of electrons orbiting each atom's nucleus." http://science.howstuffworks.com/light7.htm

[3] "Like all elementary particles, photons are currently best explained by quantum mechanics and exhibit wave–particle duality, exhibiting properties of both waves and particles."
 http://en.wikipedia.org/wiki/Photon

[3] "A very difficult answer to understand – "Atoms emit a photon when an electron falls from a high-energy state to a low-energy state. The conditions under which this process occurs happen in two ways. According to the Cornell Center for Materials Research, electrons either absorb the energy from a photon and jump to a higher energy level or a photon collides with an electron that is already in an excited state." http://www.ask.com/question/under-what-circumstances-can-an-atom-emit-a-photon

[3] Using the latest improved scanning electron microscope [SEM], we can image the electron orbitals confirming the ball bearing shape of our atoms.   This image was published in the October 2009 *Physical Review B*, Volume 80, number 16 and reported in the Scientific American Magazine - Dec. 2009 titled: *New Microscope Reveals the Shape of Atoms.*

[4] "-Protons attract electrons - "An atom's electrons have a negative electric charge; its protons are positively charged, and its neutrons have no charge. Like charges (for example, two protons) repel each other. Unlike charges (an electron and a proton) attract. In atoms, electrons and protons are held together by their opposite electric charges." http://www.polywellnuclearfusion.com/ PolywellReactor/ElectrMagn.html

[4] "-Atom's forces prevents electrons from reaching protons … It's a case of strange, but true."   http://thinkingscifi.wordpress.com/2013/06/06/electrons-protons/

[5] "In particle physics, the strong interaction (also called the strong force, strong nuclear force, or colour force) is one of the four fundamental interactions of nature, …
"The strong interaction is observable in two areas: On the larger scale, it is the force that binds protons and neutrons together to form the nucleus of an atom. On the smaller scale, it is also the force that holds quarks and gluons together to form the proton, the neutron and other particles." www.princeton.edu/~achaney/tmve/wiki100k/docs/Strong_interaction.html

[6] "… and neutrons, which carry no charge."   http://aether.lbl.gov/elements/stellar/strong/strong.html

[6] "…   and so we did experiments hitting the protons and neutrons together with higher energy   …   we got over 400 different kinds of particles   …"   Richard Feynman, *The Pleasure of Finding Things Out:,* Basic books, 2005, p. 237.

[6] "The nature of quark confinement suggests that the quarks are surrounded by a cloud of gluons, and within the tiny volume of the proton other quark-antiquark pairs can be produced and then annihilated without changing the net external appearance of the proton."   http://hyperphysics.phy-astr.gsu.edu/hbase/particles/proton.html

Figure 34: Quarks, antiquarks and gluons – letters are randomly colored red, blue, green.

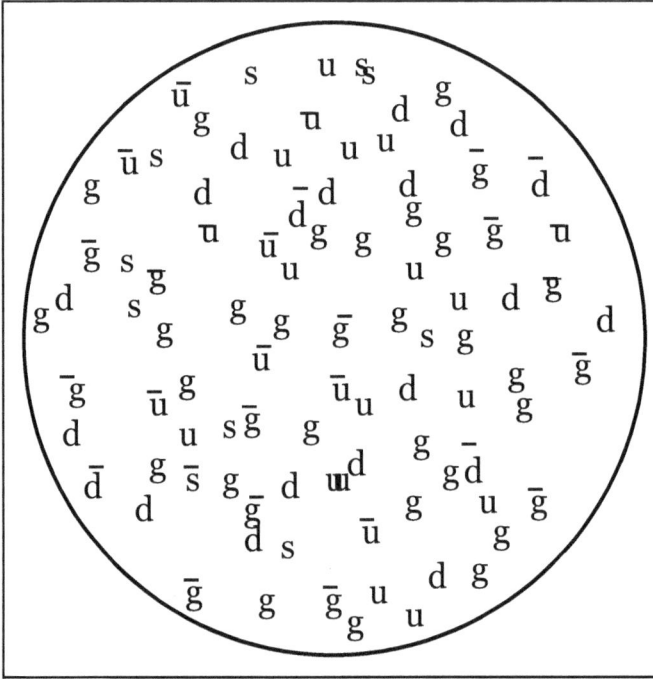

Above figure – "Snapshot of a proton - and imagine all of the quarks (up, down, and strange - u,d,s), antiquarks (u,d,s with a bar on top), and gluons (g) zipping around near the speed of light, banging into each other, and appearing and disappearing.
…

"It is impossible to describe the proton's structure simply, or draw simple pictures, because it's highly disorganized. All the quarks and antiquarks and gluons inside are rushing around as fast as possible, at nearly the speed of light.

"You may have heard that a proton is made from three quarks. Indeed *here* are several *pages* that *say* so. This is a lie — a white lie, but a *big* one. In fact there are zillions of gluons, antiquarks, and quarks in a proton."   http://profmatt strassler.com/articles-and-posts/largehadroncolliderfaq/whats-a-proton-anyway/

[6] "the quarks are considered to be held together by the color force," http://hyperphysics.phy-astr.gsu.edu/hbase/forces/funfor.html

[7] "Physicists call the theoretical framework that describes the interactions between elementary building blocks (quarks and leptons) and the force carriers (bosons) the Standard Model." http://www.fnal.gov/pub/science/inquir ing/matter/madeof/

[8] "Spin is a bizarre physical quantity. ... Based on the known sizes of subatomic particles, however, the surfaces of charged particles would have to be moving faster than the speed of light in order to produce the measured magnetic moments. ... that make spin one of the more challenging aspects of quantum mechanics." ... "In a broader sense, spin is an essential property influencing the ordering of electrons and nuclei in atoms & molecules," http://www.scientificamerican.com/article/what-exactly-is-the-spin/

[8] "Some of the basic properties of a simple proton include mass, a positive electric charge and spin. Granted, a proton does not have a very large electric charge, but it does spin very fast and, therefore, does produce a small, but noticeable, magnetic field." http://www.simplyphysics.com/page2_1.html

[9] "Metallic bonding is the principal force holding together the atoms of a metal. A metallic bond results from the sharing of a variable number of electrons by a variable number of atoms." http://chemistry.tutorvista.com/physical-chemistry/metallic-bonding.html

[9] "The bonding of adjacent atoms is an electronic process. Strong 'primary' bonds are formed when electrons in outer shells are actually transferred or shared between atoms;" http://www.ami.ac.uk/courses/topics/0204_aab/index.html

The fuel of stars - Fission and Fusion

[10] "Fission and fusion nuclear reactions are chain reactions. … "In nature, fusion occurs in stars, such as the sun." www .diffen.com/difference/Nuclear_Fission_vs_Nuclear_Fusion

[10] "of this mass only 0.6 g (0.021 oz) was transformed into a different type of energy (initially kinetic energy, then heat and light). " http://en.wikipedia.org/wiki/LittleBoy Reference & https://www.physicsforums.com/threads/hiroshima-little-boy-atomic-bomb-physics-help.587096/

[10] "In the atomic bomb that destroyed Hiroshima only 600 milligrams of uranium (less than the weight of a dime) was converted to energy, but it released the same amount of power as at least 13,000 tons of the conventional chemical explosive TNT."   http://www.unmuseum.org/buildabomb.htm

[10] **Nuclear fission: Basics** – "When a nucleus fissions, it splits into several smaller fragments. These fragments, or fission products, are about equal to half the original mass. Two or three neutrons are also emitted. The sum of the masses of these fragments is less than the original mass. This 'missing' mass (about 0.1 percent of the original mass) has been converted into energy according to Einstein's equation." http://www.atomicarchive.com/Fission/Fission1.shtml

Conclusion – our Amazing Atoms:

Electrons orbit atoms at the speed of light driven by 'dark energy' evidenced by emission of photons when activated.

There are many complicated super powerful atomic forces balanced for such a infinitely long multi-billion year old life.

A. 13 – Scanning Tunneling Microscope

"A microscope that scans the surface of a sample with a beam of electrons, causing a narrow channel of tunneling electrons to flow between the sample and the beam, and producing three-dimensional images of atomic topography and structure." http://www.thefreedictionary.com/scanning+tunneling+microscope

"Gerd Binnig, along with his colleague, Heinrich Rohrer, was awarded the Nobel Prize in Physics in 1986 for his work in scanning tunneling microscopy…which can form an image of individual atoms on a metal or semiconductor surface by scanning the tip of a needle over the surface at a height of only a few atomic diameters." http://inventors.about.com/library/inventors/blstm.htm

"scanning tunneling microscope (STM), type of microscope whose principle of operation is based on the quantum mechanical phenomenon known as tunneling, in which the wavelike properties of electrons permit them to "tunnel" beyond the surface of a solid into regions of space that are forbidden to them under the rules of classical physics." http://www.britannica.com/EBchecked/topic/526582/scanning-tunneling-microscope-STM

"atoms really look like those textbook images."
Scientific American: *New Microscope Reveals the Shape of Atoms; Improved field-emission microscope images electron orbitals, confirming their theoretical shapes*, Nov 18, 2009 By Davide Castelvecchi: "Researchers have now managed to image the electron orbitals and show for the first time that, in a sense, atoms really look like those textbook images." [Like ball bearings.] http://www.scientificamerican.com/article/the-shape-of-atoms/

A. 14 – Supernova Exploding Stars

"The Supernova Cosmology Project and the High-Z Supernova Search Team, the two international groups of astronomers and physicists who discovered the accelerating expansion of the universe, use Type Ia supernovae as 'standard candles' to measure cosmological parameters. Type Ia spectra and light curves (their rising and falling brightness over time) are all nearly alike, and they are bright enough to be seen at very great distances." http://www2.lbl.gov/Science-Articles/Archive/oldest-1a-supernova.html

"A supernova is an explosion of a massive supergiant star. It may shine with the brightness of 10 billion suns!" … "Type Ia supernovae have become very important as the most reliable distance measurement at cosmological distances, useful at distances in excess of 1000 Mpc." [Megaparsec] http://hyperphysics.phy-astr.gsu.edu/hbase/astro/snovcn.html [1,000 Megaparsecs = 3.26 billion light years]

"Supernovae, which occur within a galaxy about every 100 years, are among the brightest events in the sky. When a star explodes, it releases so much energy that it can briefly outshine all the stars in its galaxy. … When the white dwarf reaches 1.4 solar masses, or about 40 percent more massive than our Sun, a nuclear chain reaction occurs, causing the white dwarf to explode. The resulting light is 5 billion times brighter than the Sun. … "Because the chain reaction always happens in the same way, and at the same mass, the brightness of these Type Ia super- novae are also always the same." http://hubblesite.org/hubble_discoveries/dark_energy/de-type_ia_supernovae.php

"In the last decade Dark Energy has joined Dark Matter as one of the most pressing issues in physics and astrophysics today. Key in the discovery of Dark Energy, the unknown source causing the acceleration of the Universe's expansion, has been Type Ia Supernovae (SNe Ia)." http://www.pha.jhu.edu/~bfalck/SeminarPres.html

"Type Ia supernovae – ... – begin with a pair of stars orbiting one another. At least one of the pair will be a white dwarf (the dense remnant of a star the size of our own Sun after it has exhausted all its nuclear fuel) whilst the other will usually be a red giant.
"The denser white dwarf will then begin pulling matter away from its larger neighbor, and when it has absorbed enough nuclear fusion will reignite inside its core and it will explode." http://www.independent.co.uk/news/science/white-dwarf-explodes-in-the-galaxy-next-door-type-1a-supernova-spotted-in-galaxy-m82-9083909.html

A. 15: Gamma Ray Bursts - many open questions

many open questions
Gamma Ray Bursts Workshop at International Space Science Institute – Beijing: *Gamma Ray Bursts: a tool to explore the young Universe,* April 13 – 17, 2015:
"Despite the recent progresses in Gamma-Ray Burst (GRB) science, obtained in particular thanks to the Swift and Fermi satellites, there are still many open questions in the field. One concerns the mechanisms that power these extreme explosions (in a handful of seconds the isotropic equivalent energy emitted by GRBs spans from 1050 to 1054 erg, making them the most luminous events in the Universe), which is still unclear after more than four decades since their discovery."
http://www.issibj.ac.cn/Program/Workshops/GRB/201409/t20140909_127462.html

"Gamma-ray bursts (GRBs) are short-lived bursts of gamma-ray light, the most energetic form of light. Lasting anywhere from a few milliseconds to several minutes, GRBs shine hundreds of times brighter than a typical supernova and about a million trillion times as bright as the Sun. When a GRB erupts, it is briefly the brightest source of cosmic gamma-ray photons in the observable Universe." http://imagine.gsfc.nasa.gov/docs/science/know_11/bursts.html

How do gamma-ray bursts work? by Gemma Lavender**,** December 5, 2013 "That's one question that astronomers are asking after our current understanding of these bright explosions is challenged by new observations. 'We thought the visible light for these flashes came from internal shocks, but this burst shows that it must come from the external shock, which produces the most energetic gamma rays,' says Fermi team member Sylvia Zhu at the University of Maryland." www.spaceanswers.com/news/how-do-gamma-ray-bursts-work/

"Early during the eruption of the burst [GRB 130427A], a trio of NASA satellites including Swift, working in concert with ground-based robotic telescopes, began capturing never-before-seen details that challenge current theoretical under-standings of how gamma-ray bursts work." http://science.psu edu/news-and-events/2013-news/ Burrows11-2013

Astronomers View Cosmic Blast GRB 130427A in Unique Detail "Gamma-ray bursts are the most luminous explosions in the cosmos, thought to be triggered when the core of a massive star runs out of nuclear fuel, collapses under its own weight, and forms a black hole. The black hole then drives jets of particles that drill all the way through the collapsing star and erupt into space at nearly the speed of light." … "The spectacular results from Fermi GBM show that our widely accepted picture of MeV gamma rays from internal shock waves is woefully inadequate," said Rob Preece, a Fermi team member at the University of Alabama in Huntsville who led the GBM study." http://scitechdaily.com/ astronomers-view-cosmic-blast-grb-130427a-unique-detail/

"A gamma-ray burst known as GRB 090429B for the 29 April 2009 date when it was detected by NASA's Swift satellite has been found to be a candidate for the most distant object in the Universe at an estimated distance of 13.14 billion light years." http://www.dailygalaxy.com/my_weblog/2013/10/a-pale-red-dot-the-most-distant-object-in-the-universe.html

"No two gamma-ray bursts are the same, as can be seen from this sample of a dozen light curves. Some are short, some are long, some are weak, some are strong, some have more spikes, some have none, each unlike the other one." https://heasarc .gsfc.nasa.gov/docs/objects/grbs/grb_profiles.html

"They occur approximately once per day and are brief, but intense, flashes of gamma radiation." http://swift.gsfc.nasa.gov/

A. 16 CMB – Cosmic Microwave Background 'ember' 'speckle' Radiation & the Axis of Evil

"The cosmic expansion poses other puzzles. The geometry of space is very close to flat. As we saw when discussing the microwave radiation [CMB] left over from the big bang, the characteristic 'speckles' haven't been magnified or demagnified in their long travel through space, and that means space is flat to 1 percent." Chris Impey, *How It Began: A Time-Traveler's Guide To The Universe*, W. W. Norton & Company, New York, 2012, page 307.

CMB embers are undrifting 'speckles'
If the CMB was truly the light from the Big Bang Explosion reflected then NASA would have observed shifting sideways of these '**speckles**,' no shifting noted over the nine year study.

WMAP 9-year Results Released "The WMAP science team has determined, to a high degree of accuracy and precision, not only the age of the universe, but also the density of atoms; the density of all other non-atomic matter … "WMAP's 'baby picture of the universe' maps the afterglow of the hot, young universe at a time when it was only 375,000 years old, when it was a tiny fraction of its current age of 13.77 billion years. … "The first results were issued in February 2003, with major updates in 2005, 2007, 2009, 2011, and now this final release." http://map.gsfc.nasa.gov/news

The best explanation for these unchanging **speckle** positions is that they are outrushing embers from the Big Bang

Second is that these first outrushing ember **speckles** are all flying away from the Big Bang Epicenter and consequently thin out the further they get, now some 13 billion light years from Earth. But a nine year observation is insufficient for a change in brightness. – Further, these **speckles** out at 13 billion light years are not dense enough to form galaxies but like a smoke ring shot out of a cannon is too dense to see thru surrounding our observable Universe.

"Scientists using NASA's Wilkinson Microwave Anisotropy Probe (WMAP), during a sweeping 12-month observation of the entire sky, captured the new cosmic portrait, capturing the afterglow of the big bang, called the cosmic microwave background. ... "The light we see today, as the cosmic microwave background, has traveled over 13 billion years to reach us. Within this light are infinitesimal patterns [but no side motion of 'speckles'] that mark the seeds of what later grew into clusters of galaxies and the vast structure we see all around us. ... Launched on June 30, 2001, WMAP maintains a distant orbit about the second Lagrange Point, or 'L2,' a million miles from Earth." http://map.gsfc.nasa .gov/news/PressRelease_03-064.html

"The universe much beyond a redshift of 1000 is unobservable. We cannot see through the fog, as there is a photon barrier or last scattering surface at that redshift. The microwave photons that comprise the cosmic background radiation come from that last scattering surface." Jeremiah P. Ostriker and Simon Mitton, *Heart of Darkness: Unraveling the Mysteries of the Invisible Universe*, Princeton University Press, Princeton, 2013, page 153.

"The Big Bang theory predicts that the early universe was a very hot place and that as it expands, the gas within it cools. Thus the universe should be filled with radiation that is literally the remnant heat left over from the Big Bang, called the 'cosmic microwave background', or CMB." ... "Today, the CMB radiation is very cold, only 2.725° above absolute zero, thus this radiation shines primarily in the microwave portion of the electromagnetic spectrum, and is invisible to the naked eye. However, it fills the universe and can be detected everywhere we look." … "The temperature is uniform to better than one part in a [hundred thousand] thousand! [Like sand **speckles** on a beach.] This uniformity is one compelling reason to interpret the radiation as remnant heat from the Big Bang; it would be very difficult to imagine a local source of radiation that was this uniform. In fact, many scientists have tried to devise alternative explanations for the source of this radiation, but none have succeeded."[**up until this book**] http://map .gsfc.**nasa**.gov/universe/bb_tests_cmb.html

Erik M. Leitch of the University of Chicago explains. *What is the cosmic microwave background radiation?* November 1, 2004: "The Cosmic Microwave Background radiation, or CMB for short, is a faint glow of light that fills the universe, falling on Earth from every direction with nearly uniform intensity. It is the residual heat of creation--the afterglow of the big bang--streaming through space these last 14 billion years like the heat from a sun-warmed rock, reradiated at night."   http://www.scientificamerican.com/article/what-is-the-cosmic-microw/

"… the temperature variations **[speckles]** in the CMB appear to be sized and spaced differently when Planck looks in one direction, than when it looks in the other.   There are other anomalies as well. The variations don't appear to behave the same on large scales as they do on small scales, and there are some particularly large features, such as a hefty cold spot, that were not predicted by basic inflation models. Ultimately, the data show 'some features that are surprising and very, very intriguing,' said Charles Lawrence, U.S. Planck project scientist at NASA's Jet Propulsion Laboratory in Pasadena, Calif.   'Hopefully in the process of understanding those features we will be able to glimpse answers to some of our deepest questions.'" http://www.space.com/20338-big-bang-light-exotic-physics.html

cosmology may need a rethink
"Planck's new image of the CMB suggests that some aspects of the standard model of cosmology may need a rethink, raising the possibility that the fabric of the cosmos, on the largest scales of the observable Universe, might be more complex than we think." http://sci.esa.int/planck/51551-simple-but-challenging-the-universe-according-to-planck/

standard cosmological model might have to be modified
"Although the differences were slight, cosmologist David Spergel of Princeton University in New Jersey was intrigued. "Planck [regarding the CMB] is so precise that even small discrepancies **[speckles]** become interesting," he notes. The initial findings, he says, suggested one of three possibilities: Either the standard cosmological model might have to be modified; or a host of different astronomical studies were incorrect; or some systematic error in the Planck data had not been accounted for." http://www.nature.com/news/cosmo logists-at-odds-over-mysterious-anomalies-in-data-from-early-universe-1.14368

Planck CMB Anomalies: Astrophysical and Cosmological Secondary Effects and the Curse of Masking (Submitted on 8 May 2014 (v1), last revised Aug. 3 2014 (this version, v2)):
"Large-scale anomalies have been reported in CMB data with both WMAP and Planck data. These could be due to foreground residuals and or systematic effects, though their confirmation with Planck data suggests they are not due to a problem in the WMAP or Planck pipelines. If these anomalies are in fact primordial, then understanding their origin is fundamental to either validate the standard model of cosmology or to explore new physics." http://arxiv.org/abs/1405.1844

"divides the universe … into two distinct sections"
"WMAP [re the CMB] (and the newer 'Planck' Observatory) found various things that can't be explained using conventional models. Several of which are, [1st] THE CMBR COLD SPOT … [2nd] The CMBR data revealed that more than one hundred galaxy clusters are not only lit up by hot, x-ray emitting gases, but are also speeding off in the same direction, … The clusters seem to be traveling more than 2 million miles per hour, into an expanse of about 20-degrees of sky. Furthermore, the trend is not a statistical fluke, as it continues to hold steady throughout interstellar space instead of bucking black to normal speeds and distributions. Once again, our theories say this should be impossible. … [3rd] NASA's WMAP showed us that macroscopically, there is an asymmetrical pattern that divides the universe (at least according to the CMB) into two distinct sections, with each side having a different temperature relative to the other. …Of all the anomalies, this one incited the most controversy, with many physicists believing either the axis of evil didn't really exist, or that it could be explained in an extremely simple manner …" http://www.fromquarkstoquasars.com/4-anomalies-in-the-big-bang-afterglow/

The CMB's 'Axis of Evil'
The axis of evil finds that the poles of the CMB lines up with the direction of our Milky Way Galaxy contrary to what cosmology assumes that there should be no preferred direction in their 'assumed centerless' Universe.

'Axis of evil' warps cosmic background, by Marcus Chown: "Dubbed the 'axis of evil' by cosmologist João Magueijo of Imperial College London, the pattern appears in the map of the microwave background (CMB) built up by NASA's Wilkinson Microwave Anisotropy Probe (WMAP). As part of their analysis, astronomers break up the subtle temperature variations in the CMB into components called the dipole, the

quadrupole and the octupole (see Graphic), like breaking up an orchestral score into tunes played by different instruments. If the CMB really is the afterglow of the big bang, then the orientations of the hot and cold regions of the quadrupole and the octupole should be random. 'But they are not,' says Vale. 'The big surprise is they are aligned - along the axis of evil.'"
http://www.newscientist.com/article/dn8193-axis-of-evil-warps-cosmic-background.html

Axis of Evil – unexpected correlations ... with the ... direction of the solar system

Mysteries at Universe's Largest Observable Scales by Dragan Huterer November 24, 2005: "Cosmological principle states that the universe is homogeneous and isotropic on its largest scales. ... Extraordinary full-sky maps produced by the WMAP experiment, in particular, are revolutionizing our ability to test the isotropy of the universe on its largest scales. ... KICP fellow Dragan Huterer and his collaborators proposed a useful new basis for representing the CMB anisotropy, and furthermore found statistically significant and completely unexpected correlations [the axis of evil] of the CMB quadrupole and octupole with the geometry and direction of motion of the solar system." http://cfcp.uchicago.edu/research/highlights/highlight_2005-11-24.html

A.17 – Space Expansion/Acceleration, The Cosmological Constant & Quark Fog

See also 28. Inflation, Superstrings, Hidden Dimensions, Parallel Universes, and the Multiverse

Space Expansion

"First of all, *are galaxies really moving away from us, or is space just expanding?* Conveniently, Einstein's theory of gravity (general relativity) says that these are two equivalent viewpoints that are equally valid, as illustrated in Figure 3.2, so you're free to think about it in whichever way you find more intuitive.*

* "Mathematically, the different viewpoints correspond to different choices of space coordinates, and Einstein's theory allow you to pick whichever coordinate system you want for space and time." See Max Tegmark, *Our Mathematical Universe*, Alfred A. Knopf, New York, 2014, page 47.

"When astronomers say the universe is expanding, they don't mean that space is expanding, although some of us are guilty of putting it that way. ... instead, the universe is filled with galaxies rushing away from one another, which as far as we can tell fill all space, with no center and no edge." Steven Weinberg, *Lake Views: This World and the Universe*, The Belknap Press of Harvard University Press, Cambridge, 2009, pages 35 & 36.

What does it mean when they say the universe is expanding?
"The galaxies outside of our own are moving away from us, and the ones that are farthest away are moving the fastest. This means that no matter what galaxy you happen to be in, all the other galaxies are moving away from you.

"However, the galaxies are not moving through space, they are moving in space, because space is also moving. In other words, the universe has no center; everything is moving away from everything else." http://www.loc.gov/rr/scitech/mysteries/universe.html

"You can't see it happening on Earth, but space itself is stretching. ... "If you go into the distant future, everything that we see in the universe right now will expand away from us so much that we won't be able to see it anymore," said David Schlegel of the Lawrence Berkeley National Laboratory." http://www.cnn.com/2014/04/08/tech/innovation/universe-expansion-astronomers/

"Although the universe is expanding, no one knows what if anything it expands into. In order to know, astronomers would have to observe outside the space of our universe, just as observers on the balloon would have to observe outside the skin of the balloon, which the rules of analogy forbid. So astronomers are stuck. Perhaps the universe is expanding into nothing, or into God's living room, or into some mad physicist's laboratory. Take your pick." See: Ken Croswell, *The Universe at Midnight*, The Free Press, 2001, page 74.

"A common misconception is that a mysterious force must be pushing the galaxies apart. ... some initial explosion, the 'big bang,' sent the galaxies flying away from each other, but the only force operating now between galaxies is gravitational attraction." See: A. Zee, Einstein's Universe, Oxford University Press, 1989, pages 59-60.

"We have believable evidence that the universe is expanding, the space between the galaxies opening up, and that this expansion traces back to a hot dense phase, the big bang." P.J.E. Peebles, *Principles of Physical Cosmology*, Princeton University Press, Princeton, 1993, page 3

mathematically assumed space expansion
"In the early 20th century the common worldview held that

the universe is static — more or less the same throughout eternity. Einstein expressed the general opinion in 1917 after de Sitter produced equations that could describe a universe that was expanding, a universe with a beginning. Einstein wrote him that "This circumstance irritates me." In another letter, Einstein added: 'To admit such possibilities seems senseless.'" http://www.aip.org/history/cosmology/ideas/expanding.htm

"In 1929 Edwin Hubble, working at the Carnegie Observatories in Pasadena, California, measured the redshifts of a number of distant galaxies. He also measured their relative distances by measuring the apparent brightness of a class of variable stars called Cepheids in each galaxy. When he plotted redshift against relative distance, he found that the redshift of distant galaxies increased as a linear function of their distance. The only explanation for this observation is that the universe was expanding." http://skyserver.sdss.org/dr1/en/astro/universe/universe.asp [see also redshift at A25 p. 173]

Accelerating Expansion Assumption explained using Einstein's mathematical Cosmological Constant

"Perhaps because cosmic acceleration seems so exotic, theorists tend to offer exotic explanations by invoking new or unknown ingredients (exotic particles) or physical conditions (cosmic strings and leaky gravity)." See: J. Kanipe, *Chasing Hubble's Shadows,* Hill and Wang, 2006, page 103.

"Hence the faintness of the supernovae at $z = 1$ implies an accelerating expansion. Based on this data the old idea of a cosmological constant is making a comeback." http://math.ucr.edu/home/baez/physics/Relativity/GR/cos_constant.html

"The *cosmological constant* is now being used to explain the acceleration of the expansion of the universe instead of it slowing down." http://planetfacts.org/cosmological-constant/

"Hubble's second revolutionary discovery was based on comparing his measurements of the Cepheid-based galaxy distance determinations with measurements of the relative velocities of these galaxies. He showed that more distant galaxies were moving away from us more rapidly:" http://map .gsfc.nasa.gov/universe/uni_expansion.html

"While in general galaxies follow the smooth expansion, the more distant ones moving faster away from us, other motions cause slight deviations from the line predicted by Hubble's Law. This diagram [fig.16, chapter 3, page 51] shows a typical plot of distance versus recessional velocity, with each point showing the relationship for an individual galaxy." http://www .astro.cornell.edu/academics/courses/astro201/hubbles_law.htm

In reality what Hubble found was that the distances between galaxies was increasing — the rapid bizarre concept created at that time concluded that space was expanding – probably due to de Sitter's 1915 mathematical analysis that space without matter could expand.

Right Again, Einstein! New Study Supports 'Cosmological Constant,' by Clara Moskowitz, "Dark energy is the name given to whatever is causing the expansion of the universe to accelerate. One theory predicts that an unchanging entity pervading space called the cosmological constant, originally suggested by Albert Einstein, is behind dark energy." http:// www.space.com/19282-einstein-cosmological-constant-dark-energy.html

"The favored explanation for the cosmic acceleration is dark energy, a hypothetical form of energy that permeates all space and exerts a negative pressure, so as the Universe expands, the pressure increases and causes the universe to expand at an ever-increasing rate." http://chandra.harvard.edu/xray_astro /dark_energy/index2.html

"The consensus among cosmologists is that the Universe is accelerating, but this is inferred from a model of the expansion history and an unproven assumption about the uniformity of the Universe." Published July 24, 2014 http://physics. aps.org/synopsis-for/10.1103/PhysRevLett. 113.041303

"The evidence for an accelerating expansion comes from observations of the brightness of distant supernovae." http:// www.astro.ucla.edu/~wright/cosmology_faq.html#CC

nowhere seen in terrestrial laboratories –
an important problem in basic physics was not yet solved
"The idea that the universe was composed principally of vacuum energy with negative pressure, required by these astronomical observations but nowhere seen in terrestrial laboratories, meant that an important problem in basic physics was not yet solved." Robert P. Kirshner, *The Extravagant Universe: exploding stars, dark energy, and the accelerating cosmos*, Princeton University Press, Princeton, 2002, p. 236.

"So far there is only indirect evidence that the Universe is undergoing an accelerated expansion. The evidence for cosmic acceleration is based on the observation of different objects at different distances and requires invoking the Copernican cosmological principle and Einstein's equations of motion." http://journals.aps.org/prl/abstract/10.1103/PhysRevLett.113.0 41303

"Einstein admitted in his 1917 paper [*Cosmological Considerations on the General Theory of Relativity* – last paragraph] that the introduction of the cosmological constant 'is not justified by our actual knowledge of gravitation,' i. e., that it had an ad hoc character, but he found it 'necessary for the purpose of making a quasi-static distribution of matter." See: Helge Kragh, *Cosmology and Controversy,* Princeton University Press, 1996, page 9.

Einstein's math assumption – the cosmological constant

"The trouble with the cosmological constant, expressed by the Greek letter lambda, [Λ] is that no one knows if it exists at all. Einstein conceived of the concept to make his general theory of relativity conform to what was then wrongly believed to be a static universe that is not expanding." See: John Noble Wilford, *Cosmic Dispatches,* W. W. Norton, 2001, p. 216-7.

"In *The origin of the Universe,* published by Basic Books, Dr John D. Barrow, an astronomer at the University of Sussex in England, wrote: "One must understand that the term 'Big-Bang model' has come to mean nothing more than a picture of an expanding universe in which the past was hotter and denser than the present. The job of cosmologists is to pin down the expansion history of the universe–to determine how the galaxies formed; why they cluster as they do; why the expansion proceeds at the rate that it does–and to explain the shape of the universe and the balance of matter and radiation existing within it." See: John Noble Wilford, *Cosmic Dispatches,* W. W. Norton & Co., 2001, page 226.

"There was never any physical argument in favor of the cosmological constant, only the philosophical desire to make a static universe." See: David Lindley, *The End of Physics*, Basic Books, 1993, page 138.

"To make his equations work, in 1917 Einstein introduced an additional term into them, expressed by the Greek letter lambda (Λ) -- the "cosmological constant." The new term represented a repulsive force that would counter gravity's attraction, leaving the universe intact.
"But in the years that followed, evidence mounted that the belief in the universe's motionlessness was wrong: The universe was, in fact, expanding." http://www.theatlantic.com/technology/archive/2013/08/einstein-likely-never-said-one-of-his-most-oft-quoted-phrases/278508/

"To put it mildly, the cosmological constant is weird stuff. Nobody has any idea what it is, but if it exists, it is a manifestation of the energy of empty space. See: Terence Dickinson, *The Universe and Beyond*, A Firefly Book, 1999, page 121.

Quark Fog   See figure 26, page 64 See also A25 page 176 for expanded entry of this next quote – "What is the redshift of the CMB surface?   "It means that the amount of space between us and that point is growing (because of expansion) at a rate such that in one second, the amount of distance increases by 299,791.9 km (or 99.9998% of the speed of light)[in miles 186,281.6+ see also figure 26 quark fog, page 64]"   https://www.google.com/?gws_rd=ssl#q=What+is+the+redshift+of+the+CMB+surface%3F see https://answers.yahoo.com /question/index?qid=20100509112 656AA8E0FO

"For one thing, when they attempt to calculate the value of lambda [cosmological constant],   the theorists come up with a figure that is 120 orders of magnitude too big. Not 120 times too big— 10^{120} times too big. Fitting the known universe with a vacuum energy of that potency would be like filling up a water balloon with a fire hose.   … 'It cannot possibly be correct,' says Turner. 'If it were correct, you wouldn't be able to see beyond the end of your nose, the universe would be expanding so fast.'   The size of the error has emphasized how poorly physicists understand certain aspects of gravity. 'That is the biggest embarrassment in theoretical physics,' adds Turner. *Very Dark Energy,* Discover Magazine, March, 2001, Karen Wright,   http://discovermagazine.com/2001/mar/feat dark/

Also by Michael Turner –

"Complicating matters further, the repulsive force [that cosmological constant] had to be less when the galaxies were forming than it is today. If it had always hovered around its present strength, galaxies would have blown apart before they could have formed, and no one would be here today to worry about it. 'We want it to be here today and gone yesterday, so that it doesn't interfere with the growth of structure, said [Michael] Turner." See: K. C. Cole, *The Hole in the Universe: How Scientists Peered over the Edge of Emptiness and Found Everything.* New York: Harcourt, Inc., 2001, page 201.

"The expansion of the universe is nothing but a continual swelling of space. Every day the region of the universe accessible to our telescopes swells by 10^{18} cubic light-years. Where is all this space 'coming from'?" See: John Leslie, Editor, *Modern Cosmology & Philosophy*, Prometheus Books, Amherst, N.Y., 1998, page 244.

"When we examine all these measurements of the intensity of radiation coming to us from different directions in the sky, we learn a number of striking things about the structure of the universe. We find that it is expanding at the same rate in every direction to an accuracy better than one part in a thousand. We say that the expansion is isotropic—that is, the same in every direction." See: John D. Barrow, *The Origin of the Universe,* Basic Books, New York, 1994, page 16.

"But remember that the energy of empty space is gravitationally repulsive. If it had come to dominate the energy of the universe before the time of gravity formation, the repulsive force due to this energy would have outweighed (literally) the normal attractive gravitational force that caused matter to clump together. And galaxies would never have formed!" See: Lawrence M. Krauss, *A Universe From Nothing: Why There Is Something Rather than Nothing.* New York: Free Press, 2012, page 125.

A. 18 Dark energy

"In the 1990s, scientists studying exploding stars called supernovae in far-flung galaxies discovered that the Universe's expansion is accelerating, not slowing as theorists predicted. This discovery led them to the conclusion that some unknown process was causing the Universe to speed up, and they named it dark energy." http://www.bbc.co.uk/science/space/universe/ questions_and_ideas/dark_energy

"a theoretical repulsive force that counteracts gravity and causes the universe to expand at an accelerating rate." https://prezi.com/1pqnsdzgi8au/what-is-dark-energy/

Prospects for probing the dark energy via supernova distance measurements (Submitted on 13 Aug 1998 (v1), last revised 24 Jun 1999 (this version, v2)) Dragan Huterer, Michael S. Turner (Chicago/Fermilab): "Distance measurements to Type Ia supernovae (SNe Ia) indicate that the Universe is accelerating and that two-thirds of the critical energy density exists in a dark-energy component with negative pressure. Distance measurements to SNe Ia can be used to distinguish between different possibilities for the dark energy,..." http ://arxiv.org/abs/astro-ph/9808133

current theory cannot explain the acceleration
"Dark energy, on the other hand, originates from our efforts to understand the observed accelerated expansion of the universe. In a nutshell, current theory cannot explain the acceleration." www.scientificamerican.com/article/what-are-dark-matter-and/

Inferred
"This value of dark energy was inferred from a study of the expansion of the universe, as indicated by way that distant galaxies are rushing away from us." Steven Weinberg, *Lake*

Views: This World and the Universe, The Belknap Press of Harvard University Press, Cambridge, 2009, p 50.

" … but they have hardly a clue about dark energy. In 2003, the National Research Council listed 'What Is the Nature of Dark Energy?' as one of the most pressing scientific problems of the coming decades. The head of the committee that wrote the report, University of Chicago cosmologist Michael S. Turner, goes further and ranks dark energy as 'the most profound mystery in all of science.'" http://www.smithsonian mag.com/science-nature/dark-energy-the-biggest-mystery-in-the-universe

"Dark energy, which also goes by the names of the cosmological constant or quintessence, must exist due to the rate of expansion we observe for our universe. Not only is the universe expanding, but this expansion is also accelerating so the unknown 'anti-gravity' force at work is termed 'dark energy'." http://curious.astro.cornell.edu/about-us/108-the-universe/cosmology-and-the-big-bang/dark-matter/658-what-s-the-difference-between-dark-matter-and-dark-energy-intermediate

dark energy … is a complete mystery
"More is unknown than is known. We know how much dark energy there is because we know how it affects the Universe's expansion. Other than that, it is a complete mystery. But it is an important mystery. It turns out that roughly 68% of the Universe is dark energy." http://science.nasa.gov/astrophysics/focus-areas/what-is-dark-energy/

" … a group of cosmologists and physicists from Princeton University and Lawrence Berkeley National Laboratory survey the wide range of evidence which, they write, "is forcing us to consider the possibility that some cosmic dark energy exists that opposes the self-attraction of matter and causes the expansion of the universe to accelerate. ...

"'The universe is made mostly of dark matter and dark energy,' says Saul Perlmutter, leader of the Supernova Cosmology Project headquartered at Berkeley Lab, 'and we don't know what either of them is.' He credits University of Chicago cosmologist Michael Turner with coining the phrase 'dark energy' in an article they wrote together with Martin White of the University of Illinois for *Physical Review Letters*." http://www2.lbl.gov/Science-Articles/Archive/dark-energy.html

"Of course, we don't understand that picture"

" 'We have an amazingly simple picture of the universe,' says Princeton University astrophysicist Michael Strauss. 'Of course, we don't understand that picture—we don't know what dark energy is, and we don't know what dark matter is.'" http://www.scientificamerican.com/article/geometric-test-universe/

A 19. Blue-Shifted Galaxies – Moving towards Earth

"There are about 100 known galaxies with blueshifts out of the billions of galaxies in the observable universe." … the blue shifted galaxies ([are] moving towards us)." http://www .physlink.com/Education/AskExperts/ae384.cfm

"Andromeda is not the only galaxy to be moving towards us. With the help of galaxy surveys, astronomers have found that around 100 galaxies are moving towards us. Compared to the numbers of galaxies that we know of (hundreds of billions), blue-shifted galaxies are seemingly quite rare. Those that are moving towards us are either part of our Local Group, which means that we are gravitationally connected to each other, or they are found in the Virgo Cluster which everything in our Local Group is moving towards. The galaxies M90, M86 and M98 are all in the Virgo Cluster and all show blue shifts." http://www.spaceanswers.com/deep-space/2191/apart-from-andromeda-are-any-other-galaxies-moving-towards-us/

Messier Monday: The Most Blueshifted Messier Object, M86: "This is actually fairly typical of galaxies in the Virgo cluster, to have what we call peculiar velocities of as much as a few thousand km/sec. Messier 86 isn't even the fastest blueshift in the Virgo cluster; that distinction belongs (currently) to IC 3258, which approaches us at more than 500 km/sec. But it is the most blueshifted of not only all Messier galaxies, but of all Messier objects, period!" http://scienceblogs.com/startswitha bang/2013/06/10/messier-monday-the-most-blueshifted-messier-object-m86/

"In 1969 Margaret Burbridge and M. Demoulin made the first announcement of a blushifted object near the center of the Virgo cluster, IC 3258 with a approach velocity of 517 kilometers per second ! …   [517 kilometers = 321.249 miles]

"Several blueshifted galaxies appear in the direction of the Virgo cluster.

"Are we to assume that these galaxies cast doubts on the expanding Universe theory ? Not necessarily so. Some theories about the blueshifts are:

1) The expansion of the Universe is not uniform, the Virgo cluster/cloud and the Local Group may have large individual space motions relative to the general expansion.

2) The observed V_R could be due to large random motions of the galaxies within the cluster, possibly due in part to a general rotation of the cluster.

3) These blueshifted objects could have been ejected or may be escaping from the cluster and could actually be foreground objects."

http://www.poyntsource.com/Richard/blueshifted_galaxies.htm

If space expansion was a real force of nature as discussed in chapter 5 – galaxies could not form let alone have individual motion; – there would be no blueshifted galaxies.

A. 20. Galaxy collisions

Galaxy collisions - not possible if Space Expansion was a real physical occurrence pushing all celestial objects uniformly creating quark fog [chapter 5, figure 26, page 64.] Noting that galaxy collisions are commonplace, this evidence denies the possibility of Space Expansion.

APOD May 14, 2013- *Galaxy Collisions: Simulation vs Observations Explanation:* "What happens when two galaxies collide? Although it may take over a billion years, such titanic clashes are quite common. Since galaxies are mostly empty space, no internal stars are likely to themselves collide. Rather the gravitation of each galaxy will distort or destroy the other galaxy, and the galaxies may eventually merge to form a single larger galaxy." http://apod.nasa.gov/apod/ap130514.html

"   ... Hubble observations reveal that the stars in Andromeda's halo are much younger than those in the Milky Way. From this and other evidence, astronomers infer that Andromeda has already smashed into at least one and maybe several other galaxies." http://science.nasa.gov/astrophysics/focus-areas/what-are-galaxies/

Galaxy Collisions: A Preliminary Study by R. H. Miller, Dept of Astronomy, U of Chicago , reported in the Astrophysical Journal, 235:421-436, January 15, 1980:
"Galaxy collisions are extremely complicated processes in which many competing effects can only be sorted out by a method that takes all these effects into account. All of the features that make studies in galactic dynamics difficult — nonlinearity, long-range interactions, collective effects, and so on — enter along with new features that arise from the strong disturbances and rapidly changing conditions in a collision."
...   "At least 14 free parameters are required to specify a

galaxy collision for a dynamical study. The resulting 14-dimentional parameter space must be drastically limited to be studied by controlled numerical experiments." http://adsabs.harvard.edu/full/1980ApJ...235..421M

"NASA rocket experiment has revealed that up to half of all stars in the universe are celestial orphans, stars that have drifted away from their parent galaxy. ...
"Stars become orphans when galaxies that drift through space periodically collide with each other. During these galactic collisions, some of the stars are stripped away from the galaxies where they were born." ... http://www.techtimes.com/articles/19675/20141107/half-of-all-stars-may-be-orphans-and-existing-without-galaxies-ciber-observations-suggest.htm

" 'We think stars are being scattered out into space during galaxy collisions,' said Michael Zemcov, lead author of a new paper describing the results from the rocket project and an astronomer at the California Institute of Technology (Caltech) and NASA's Jet Propulsion Laboratory (JPL) in Pasadena, California. 'While we have previously observed cases where stars are flung from galaxies in a tidal stream, our new measurement implies this process is widespread.' " http://science.nasa.gov/science-news/science-at-nasa/2014/06nov_ciber/

"April 24, 2008: Astronomy textbooks typically present galaxies as staid, solitary, and majestic island worlds of glittering stars. But galaxies have a dynamical side. They have close encounters that sometimes end in grand mergers and overflowing sites of new star birth as the colliding galaxies morph into wondrous new shapes. Today, in celebration of the Hubble Space Telescope's 18th launch anniversary, 59 views of colliding galaxies constitute the largest collection of Hubble images ever released to the public. This new Hubble atlas dramatically illustrates how galaxy collisions produce a remarkable variety of intricate structures in never-before-seen detail." http://hubblesite.org/newscenter/archive/releases /2008/16

"An image of a galaxy collision captures only one stage of billion-year-long collision process. This visualization of a galaxy collision supercomputer simulation shows the entire collision sequence, and compares the different stages of the collision to different interacting galaxy pairs observed by NASA's Hubble Space Telescope. http://hubblesite.org/news center/archive/releases/2008/16/video/d/

If space expansion was a real force of nature as discussed in chapter 5 – galaxies could not collide let alone have individual motion.

A. 21 – Copernican/Cosmological Principle & The Hasty Generalization Fallacy

The Copernican/Cosmological Principle is an Assumption:
"The cosmological principle is nothing but an assumption about a symmetry of the universe as a whole. ... "Mathematically, we say that the universe as a whole is both translationally and rotationally symmetric."   See: Henning Genz, *Nothingness*, Perseus books, 1994, page 38.

Generalizing our tiny solar system to model Our Universe:
"The Copernican Principle extrapolates the revolutionary idea of sixteen-century astronomer Nicolaus Copernicus – that Earth is not the center of everything and that the Sun and planets do not revolve around it – to the universe as a whole." Paul Halpern, PhD, *Edge of the Universe: A Voyage to the Cosmic Horizon and Beyond*, John Wiley & Sons, 2012, p. 12.

"Modern cosmologists even speak of a Copernican principle: the rule that no cosmological theory can be taken seriously that puts our galaxy at any distinctive place in the universe." See Steven Weinberg, *Dreams of a Final Theory*. New York: Pantheon Books, 1992 p.246. [**a hasty generalization fallacy**]

"The Big Bang model of cosmology rests on two key ideas that date back to the early 20th century: General Relativity and the Cosmological Principle."   http://map.gsfc.nasa.gov/univer se/bb_concepts.html

"The simplest assumption to make is that if you viewed the contents of the universe with sufficiently poor vision, it would appear roughly the same everywhere and in every direction. That is, the matter in the universe is homogeneous and isotropic when averaged over very large scales. This is called the Cosmological Principle." http://map.gsfc.nasa.gov/unive rse /bb_theory.html   [**another hasty generalization fallacy**]

[another generalization fallacy]

"In order to make the cosmological problem tractable he [Einstein] made the dramatic proposal that on the large scale the universe is homogeneous, the same at every point, and isotropic, that is, it looks the same in every direction." See: Michael Rowan-Robinson, *The Nine Numbers of the Cosmos,* Oxford University Press, 1999, pages 28-9

"According to this [cosmological] principle, which he [Milne] first formulated in 1933, the world [Universe] must appear the same to all equivalent observers, irrespective of their positions." See: Helge Kragh, *Cosmology and Controversy,* Princeton University Press, 1996, page 62.

"The door opened by Copernicus could not be closed. We now call the idea that we are not in a special place, the Copernican principle." See: Michael Rowan-Robinson, *The Nine Numbers of the Cosmos,* Oxford U. Press, 1999, page 25.

"What we call the *cosmological principle* is the tenet that an observer placed at a random point in our universe will not be able to distinguish, by a look in any direction he chooses, this point and this direction from any other." See: Henning Genz, *Nothingness: The science of empty space.* Cambridge, Massachusetts: Perseus books, 1994, page 266.

"Application of the Theory of General Relativity to the large-scale structure of the Universe leads to various cosmological theories. ... The starting point for these theories is what is termed cosmological principle: Viewed on sufficiently large distance scales, there are no preferred directions or preferred places in the Universe.

"Stated simply, this principle means that averaged over large enough distances, one part of the Universe looks approximately like any other part." http://csep10.phys.utk.edu/astr 162/lect/cosmology/cp.html

"The Copernican Principle is a basic statement in physics that there should be no 'special' observers. For example, the Aristotelian model of the solar system in the Middle Ages placed the Earth at the center of the solar system, a unique place since it 'appears' that everything revolved around the Earth. Nicolaus Copernicus demonstrated that this view was incorrect and that the Sun was at the center of the solar system with the Earth in orbit around the Sun. ... The lesson learned by future scientists is that if a theory requires a special origin or viewpoint, then it is <u>not plausible</u>. Almost all cosmological and scientific theories are scrutinized by the Copernican principle. Often interpreted that is [is if] an idea requires some special condition, then it is incomplete."
http://abyss.uoregon.edu/~js/cosmo/lectures/lec05.html

The Hasty Generalization Fallacy
"(also known as: argument from small numbers, statistics of small numbers, insufficient statistics, unrepresentative sample, argument by generalization, hasty conclusion, inductive generalization, insufficient sample, lonely fact fallacy, over generality, over generalization)" "Drawing a conclusion based on a small sample size, rather than looking at statistics that are much more in line with the typical or average situation.
"Variation: The *hasty conclusion* is leaping to a conclusion without carefully considering the alternatives -- a tad different than drawing a conclusion from too small of a sample." http://www.logicallyfallacious.com/index.php/logical-fallacies/101-hasty-generalization

A.22 Math Based Curved Space, Universe with No Center, & Some Science considers Location of Center

See also Flat Space at A28 page 194

The Mathematically Curved Space Assumption
"Einstein developed a set of equations to describe the manner in which space-time is distorted by matter. These equations make use of a geometry developed by the 19th-century German mathematician Georg F. B. Riemann. In Riemannian geometry there are no straight lines but only curves. Therefore, the space described by the general theory of relativity is a curved space without straight lines." http:// science.howstuffworks.com/dictionary/physics-terms/theoryof -relativity-info3.htm

"This, in a nutshell, then, is the General Theory of Relativity, and its central premise is that the curvature of space-time is directly determined by the distribution of matter and energy contained within it. What complicates things, however, is that the distribution of matter and energy is in turn governed by the curvature of space, leading to a feedback loop and a lot of very complex mathematics." http://www.physicsoftheuniverse.com /topics_relativity_general.html

"The idea of general relativity is not very hard to understand. The mathematics of it is quite complicated and involves curved space geometry that is not easy to comprehend. Einstein had struggled with the mathematics of his theory for several years before he got to the correct version of his famous field equation. Though looks quite simple, this equation actually includes 10 different differential equations, and cannot be used in practice as it is." http://www.rafimoor.com/ english/GRE1.htm

Centerless Universe
No center based on the math of 1932-4, Edward A. Milne:

"In 1932-4, Edward A. Milne formulated the "*cosmological principle*" describing a homogenous universe that looks the same regardless of one's position in our Universe. According to Milne's principle, every observer in the universe should get the same world picture, that is, should make precisely the same observations of the universe at the same moment as any other observer." (Milne 1934b)".at: http://plato.stanford.edu/entries/cosmology-30s/#ButMilEnd

Singularity – "where our theories cease to be meaningful"

"On to the question, the Universe, at least according to current theory, has no centre. The important thing to realize is that a 'singularity' is not a physical thing, i.e. you can't say 'oh look at that singularity over there'. In fact a singularity is merely a situation where our theories cease to be meaningful. The Big Bang model says that the Universe was hotter and denser in the past, but current theory doesn't actually let us wind the clock back to the very beginning. The equations themselves stop being meaningful at some point."   http://www.physics forums. com/showthread.php?t=232453

"Where is the actual center of the Universe, regardless of who's observing it? Our Universe might be finite or it might be infinite. Astronomers don't actually know for sure." www.universetoday.com/111561/where-is-the-center-of-the-universe/

"Today astronomers believe that there is no centre to the cosmos. You might think that there must be a central point – after all, the Big Bang must have started somewhere? While great explosions of say, a bomb, do start from one point, the Big Bang that is believed to have created our Universe nearly 14 billion years ago was a different matter entirely and appeared to happen everywhere all at once – time and space did not exist before the Big Bang and so there was no point from where it could have erupted from."   http://www.space answers.com/deep-space/is-there-a-centre-of-the-universe/

164

"Is it possible to point to a direction in the sky and say 'that way is the center of the universe, where the Big Bang started?' … Now we know that not only are we humans not at the center of the universe, but there is no center of the universe!" http://spaceplace.nasa.gov/review/dr-marc-space/center-of-universe.html

"But the Big Bang was not an explosion IN space. It was an explosion OF space! According to the physics as we now understand it, there was no space, matter, energy, or time before the Big Bang. Because the Big Bang was everything, it happened everywhere. Observation seems to bear this out. On a very large scales, there is no motion or distribution of galaxies that suggest a center in the universe, or any privileged point where matter flows to or from." http://oneminuteastronomer.com/6949/where-is-the-center-of-the-universe/#sthash.2cjaTwq3.dpuf

"In reality, ALL of space was filled with energy right from the beginning. There was no center to the expansion, and no magical point from which matter hurtled outward." http://www.cfa.harvard.edu/seuforum/faq.htm

"When astronomers look at distant galaxies to determine how fast they're moving, it looks like they're all moving away from us. Does that mean we're at the center of the universe? Well, no. It turns out that every point in the universe sees itself as the center!" http://www.exploratorium.edu/hubble/tools/center.html

"What happens at the edge of the Universe? The short (but rather confusing) answer is that the Universe has no edges. It has no edges, no ends, and no center. All points within the Universe appear the same, and the Universe looks the same no matter where you are standing inside it. But what does this really mean?" http://astronomy.nmsu.edu/geas/lectures/lecture 28/slide02.html

Perhaps Earth is in the Center of our Universe

"In conclusion – with our current understanding, it makes as much sense to say that there is no center of the universe as it would, to say that there is. However, an interesting thing to note is that the "bubble" we call the observable universe continues to expand in each direction, giving off the impression that we are the center." http://www.fromquarkstoquasars.com/does-the-universe-have-a-center/

"If the universe possesses a center, we must be very close to it, within less than 0.1 percent of the radius; otherwise observable anisotropy in the radiation intensity would be produced, and we would detect more radiation from one direction than from the opposite direction." See: Joseph Silk, *The Big Bang*, W. H. Freeman and Company, NY, 1989, page 60. **[We do. See page 139 –"asymmetrical pattern" …in center of quote]**

"Moreover, there are fewer faint bursts [Gamma Ray Bursts] than would be expected if the sources were distributed uniformly throughout space. Apparently we are located very close to the center of a spherical arrangement of bursters." See: Leonard, P., & Bonnell, J., "gamma-ray bursts of doom," *Sky & Telescope,* (2/98), page 29.

"In 1936 the American mathematician Howard Roberson, and independently the British mathematician A. G. Walker, proved a very interesting mathematical result: if the universe looks isotropic, then either we are in a special place or the universe must also be homogeneous." See: Michael Rowan-Robinson, *The Nine Numbers of the Cosmos,* Oxford University Press, 1999, page 30.

166

A. 23 – Hubble Deep Fields, Cold Spot, and Celestial Sphere

Hubble Deep Fields

"The image, called the Hubble Deep Field (HDF), was assembled from 342 separate exposures taken with the Wide Field and Planetary Camera 2 (WFPC2) for ten consecutive days between December 18 and 28, 1995." News Release Number: STScI-1996-01 http://hubblesite.org/news center/archive/releases /1996/1996/01 /text/results/100/

"The image [Hubble Ultra Deep Field] required 800 exposures taken over the course of 400 Hubble orbits around Earth. The total amount of exposure time was 11.3 days, taken between Sept. 24, 2003 and Jan. 16, 2004." News Release Number: STScI-2004-07 http://hubblesite.org/newscenter/archive /releases /2004/07 /image/c/

"Astronomers analyzing two of the deepest views of the cosmos made with NASA's Hubble Space Telescope have uncovered a gold mine of galaxies, more than 500 that existed less than a billion years after the Big Bang. …. This sample represents the most comprehensive compilation of galaxies in the early universe, researchers said.
"The galaxies unveiled by Hubble are smaller than today's giant galaxies and very bluish in color, indicating they are ablaze with star birth. The images appear red because of the galaxies' tremendous distance from Earth. The blue light from their young stars took nearly 13 billion years to arrive at Earth. "Finding so many of these dwarf galaxies, but so few bright ones, is evidence for galaxies building up from small pieces — merging together as predicted by the hierarchical theory of galaxy formation," said astronomer Rychard Bouwens of the University of California, Santa Cruz, who led the Hubble study." STScI-2006-12 Sept. 21, 2006 http://hubblesite.org /newscenter/archive/releases/2006/12/text/

Cold Spot – the circle noted near bottom on the back cover

"Astronomers have found an enormous hole in the Universe, nearly a billion light-years across, empty of both normal matter such as stars, galaxies, and gas, and the mysterious, unseen "dark matter." While earlier studies have shown holes, or voids, in the large-scale structure of the Universe, this new discovery dwarfs them all. … "We already knew there was something different about this spot in the sky," [Lawrence] Rudnick said. The region had been dubbed the "WMAP Cold Spot," because it stood out in a map of the Cosmic Microwave Background (CMB) radiation made by the Wilkinson Microwave Anisotopy Probe (WMAP) satellite, launched by NASA in 2001." http ://www.nrao.edu/pr/2007/coldspot/

WMAP Cold Spot: "Located in Constellation Eridanus, the River, this cosmologically anomalous region was found during 2004 in a map of the Cosmic Microwave Background (CMB) radiation made by NASA's Wilkinson Microwave Anisotopy Probe (WMAP) satellite. The CMB is an imprint of microwave radiation left from about 379,000 years after the Big Bang, the hypothesized beginning of the universe, and so provides a "picture" of conditions at a very early point in time. As measured by the WMAP satellite, irregularities in the CMB show structures already present, representing temperature differences that vary by only millionths of a degree." http://www.solstation.com/x-objects/greatvoi.htm

Celestial Sphere

"Define coordinates by the projection of the Earth's pole and equator onto the celestial sphere." http://burro.case.edu /Academics/Astr306/Coords/coords.html

A. 24 Almost Instant Explosive conversion of fuel into ash with leading Spikes like CMB ember 'speckles'

Explosions are a Burning
"Subsonic explosions are created by low explosives through a slower burning process known as deflagration." http://en.wiki pedia.org/wiki/Explosion

"when combusted, coal and oil release higher levels of harmful emissions, including a higher ratio of carbon emissions, nitrogen oxides (NOx), and sulfur dioxide (SO2). Coal and fuel oil also release ash particles into the environment," http://naturalgas.org/environment/naturalgas/

Explosions are a rapid conversion of
Fuel into Ash

Rapid is not instant!
Rapid - Moving, acting, or occurring
with great speed.

"explosion - a large-scale, rapid, or spectacular expansion or bursting out or forth." www.merriam-webster.com/dictionary /explosion

"explosion - 2. (Chemistry) a violent release of energy resulting from a rapid chemical or nuclear reaction, especially one that produces a shock wave, loud noise, heat, and light." http://www.thefreedictionary.com/explosion

"Explosive: A chemical that causes a sudden, almost instantaneous release of pressure, gas, and heat when subjected to sudden shock, pressure, or high temperature." - George Mason University - Material Safety Data Sheet - See: Glossary of Common MSDS Terms.pdf

"An explosion is defined as the rapid increase in volume and release of energy in an extreme manner." http://cen acs.org /articles/91/i36 /Definition-Explosion.html

"Material that causes a sudden, almost instantaneous, release of gas, heat, and pressure, accompanied by loud noise when subjected to a certain amount of shock, pressure, or temperature." http://www.businessdictionary.com/definition/ explosive.html

"An explosion is a rapid increase in volume and release of energy in an extreme manner, usually with the generation of high temperatures and the release of gases. Supersonic explosions created by high explosives are known as detonations and travel via supersonic shock waves." http://en. Wikipedia.org/wiki/Explosion

"Using the energy release from the nuclear fission of uranium-235, an explosive device can be made by simply positioning two masses of U-235 so that they can be forced together quickly enough to form a critical mass and a rapid, uncontrolled fission chain reaction. ... the pieces must be forced together quickly and in such a geometry that the generation time for fission is extremely short. This leads to an almost instantaneous buildup of the chain reaction, creating a powerful explosion before the pieces can fly apart." http ://hyperphysics.phy-astr.gsu.edu/hbase/nucene/bomb.html

"A nuclear explosion is an explosion that occurs as a result of the rapid release of energy from a high-speed nuclear reaction." http://en.wikipedia.org/wiki/Nuclear_explosion

"In a fission reaction, a single large atom, such as uranium, is bombarded with neutrons, causing the nucleus of the atom to split into two smaller nuclei and several neutrons. The combined mass of the product atoms and neutrons is less than the mass of the original atom, and the mass loss is converted to energy according to Einstein's equation. The neutrons produced by the fission reaction cause other large atoms to fission, and their neutron production causes still other atoms to fission, leading to a chain reaction that continues exponentially (i.e., 2, 4, 8, 16, 32, 64…). The entire process is very rapid, taking only a few millionths of a second." http ://www.chemistryexplained.com/Di-Fa/Explosions.html

Explosions are not smooth – but have surface spikes –just like the speckles [a16] making up the CMB

"Using high speed cameras, molecular dynamics simulations, and back-of-the-envelope calculations we found out the key piece missing in the puzzle is a Coulomb explosion of the metal prior to the steam and hydrogen blast. As electrons leave the metal for water (to react there creating hydrogen and hydroxide) the metal charges positively to the extent that it becomes unstable (so called Rayleigh instability, same as in electrospray) and shoots out metal spikes into water, ensuring thus efficient mixing of reactants and enabling the explosive behavior."   http://marge.uochb.cas.cz/~jungwirt/

Very few studies of explosive spikes; best available is:

A 'Saturn Missile' type firework exploding underwater showing irregularities/spikes in expanding surface. View at: http://www.youtube.com/watch?v=g6ucN1Qa-Po

A.25 Spectroscopy, Redshift, Doppler Shift, Recession, & 'Z'

Spectroscopy

"The invention of spectroscopy gave chemists a powerful new tool. In many cases, the amount of an element present in a sample is too small to see. But the element is much easier to detect by spectroscopy. When the substance is heated, the [prismatic light from] hidden elements give off characteristic spectral lines. Using spectroscopy, a chemist can identify the elements by these distinctive lines." www.chemistryexplained .com/elements/A-C/Cesium.html#ixzz3P1hcmvSH

"Each element has several prominent, and many lesser, emission lines in a characteristic pattern." http://imagine.gsfc nasa.gov/docs/science/how_ll/spectral_what.html

"As you have seen, galaxies are made of stars, and stars have spectral lines, most of which are absorption lines. When a source of electromagnetic radiation is moving relative to the observer, the wavelength of the light will be redshifted if the two are moving apart and blueshifted if they are moving together." Christopher De Pree, PhD, *THE COSMOS: Idiot's Guides as Easy as It Gets!*, New York, 2014,page 202.

Redshift & Doppler Shift Effect

"In physics, redshift happens when light or other electromagnetic radiation from an object is increased in wavelength, or shifted to the red end of the spectrum. In general, whether or not the radiation is within the visible spectrum, "redder" means an increase in wavelength – equivalent to a lower frequency and a lower photon energy, in accordance with, respectively, the wave and quantum theories of light."

"Some redshifts are an example of the Doppler effect, familiar in the change in the apparent pitches of sirens and frequency of the sound waves emitted by speeding vehicles. A redshift occurs whenever a light source moves away from an observer. Another kind of redshift is cosmological redshift, which is due to the expansion of the universe, and sufficiently distant light sources (generally more than a few million light years away) show redshift corresponding to the rate of increase in their distance from Earth." http://en.wikipedia.org/wiki/Redshift

Doppler Shift Effect

"The light from distant stars and more distant galaxies is not featureless, but has distinct spectral features characteristic of the atoms in the gases around the stars. When these spectra are examined, they are found to be shifted toward the red end of the spectrum. This shift is apparently a Doppler shift and indicates that essentially all of the galaxies are moving away from us. Using the results from the nearer ones, it becomes evident that the more distant galaxies are moving away from us faster. This is the kind of result one would expect for an expanding universe." Georgia State University **[gsu]** http ://hyperphysics.phy-astr.**gsu**.edu/hbase/astro/redshf.html

"The Doppler effect can be described as the effect produced by a moving source of waves in which there is an apparent upward shift in frequency for observers towards whom the source is approaching and an apparent downward shift in frequency for observers from whom the source is receding. … "The Doppler effect can be observed for any type of wave - water wave, sound wave, light wave, etc. … [Example:] "As the car approached with its siren blasting, the pitch of the siren sound (a measure of the siren's frequency) was high; and then suddenly after the car passed by, the pitch of the siren sound was low." http://www.physicsclassroom.com/class /waves/Lesson-3/The-Doppler-Effect

Recession = the amount of space between us and a point such as a galaxy or other celestial object is increasing.

Recession
"Astronomers can also estimate the speed of recession of the distant galaxies by measuring their redshifts. ... The ubiquitousness of the shift to the red end of the spectrum, the redshift as astronomers, is due to the expansion of the universe, discovered by Edwin Hubble in the 1920's. Amir D. Aczel, *God's Equation: Einstein, Relativity, and the Expanding Universe*, Four Walls Eight Windows, NY, page 5.

" ... Hubble's law, which asserted that the recession rate of the galaxies is linearly proportional to distance. In short, galaxies move faster as they are farther away from the observer [on Earth], which also was a prediction of, and evidence for, the expanding universe theory." Joseph Silk, *Horizons of Cosmology: Exploring Worlds Seen and Unseen*, Templeton Press, West Conshohocken, Pa., 2009, page 5.

Speed of the CMB embers leaving the Big Bang Epicenter
"WMAP carried out extensive measurements of the 2.725^0 K CMB (Cosmic Microwave Background). WMAP astrophysicists estimate the Big Bang occurred 13.7 billion years ago, ... They also estimate that the CMB's redshift [z] is 1089. We can use following equation to estimate the CMB's velocity:".

$$"\frac{V}{C} = \frac{(1089 +1)^2 - 1}{(1089 +1)^2 + 1}" = 0.9999983$$

"Or, the CMB is receding from us at 0.9999983 [or 99.99% of the speed of light.]" last updated 04/16/2004 at http://www asterism.org/tutorials/tut29-1.htm

Z

"This Cosmological Calculator lets you enter values of z [redshift parameter z] and find the corresponding light travel time. This tells you the number of years the light from the object has traveled to reach us." … The table below gives light travel times and distances for some sample values of **z** . Las Cumbres Observatory Global Telescope [**lcogt**]

Z	Time the light has been traveling
0.0000715	- 1 million years
0.10	1.286 billion years
0.25	2.916 " "
0.5	5.019 " "
1	7.731 " "
2	10.324 " "
3	11.476 " "
4	12.094 " "
5	12.469 " "
6	12.716 " "
7	12.888 " "
8	13.014 " "
9	13.110 " "
10	13.184 " "

http://**lcogt**.net/spacebook/redshift

This is the outward fleeing speed of the CMB surface

"The 'speed of recession' of a point observed to be at the decoupling event, is a bit above this value of 0.999998c.

… This does NOT mean the point is moving at that speed. It means that the amount of space between us and that point is growing (because of expansion) at a rate such that in one second, the amount of distance increases by 299,791.9 km (or 99.9998% of the speed of light)" www.google.com/?gws_rd=ssl#q=What+is+the+redshift+of+the+CMB+surface%3F **OR SEE** https://answers.yahoo.com/question/index?qid= 2010050 9112 656AA8E0FO

A.26 The impossible – Space is expanding faster than Light using Math – based on an Estimate of Cosmic Inflation

Assumed Cosmic Inflation **Estimates** Size of Universe

"How far away is the edge of the universe? "Posted on January 13, 2010 by The Physicist. — Physicist: If you ever hear a physicist talking about 'the edge of the universe,' what they probably mean is 'the edge of the *visible* universe.' The oldest light (from the most distant sources) is around 15 billion years old. Through a detailed and very careful study of cosmic inflation we can estimate that those sources should now be about 45 billion light years away. So if you define the size of the visible universe as the present physical distance (in terms of the 'co-moving coordinates' which are stationary with respect to the cosmic microwave background) to the farthest things we can see, then the edge of the visible universe is 45 billion light years away (give or take)." http ://www.askamathematician.com/2010/01/q-how-far-away-is-the-edge-of-the-universe/

"Of course, the universe is only estimated to be 13.7 billion years old. And as a 'light-year' is the distance that light can travel in a year, how could the observable universe be 93 billion light-years across? Logically, light wouldn't even have time to travel more than 13.7 billion light-years.
"It seems confusing, but there is a simple explanation: the universe has expanded in all directions since the big bang."
http://www.fromquarkstoquasars.com/from-quark-to-quasar-the-observable-universe-2/

This next quote is so technical it is indecipherable:
From the UCLA .edu – Frequently Asked Questions in Cosmology – 9th question in list –

"If the Universe is only 14 billion years old, how can we see objects that are now 47 billion light years away?"

"What is the distance NOW to the most distant thing we can see? Let's take the age of the Universe to be 14 billion years. In that time light travels 14 billion light years, and some people stop here. But the distance has grown since the light traveled. The average time when the light was traveling was 7 billion years ago. For the critical density case, the scale factor for the Universe goes like the 2/3 power of the time since the Big Bang, so the Universe has grown by a factor of $2^{2/3} = 1.59$ since the midpoint of the light's trip. But the size of the Universe changes continuously, so we should divide the light's trip into short intervals. First take two intervals: 7 billion years at an average time 10.5 billion years after the Big Bang, which gives 7 billion light years that have grown by a factor of $1/(0.75)^{2/3} = 1.21$, plus another 7 billion light years at an average time 3.5 billion years after the Big Bang, which has grown by a factor of $4^{2/3} = 2.52$. Thus with 1 interval we got 1.59 x 14 = 22.3 billion light years, while with two intervals we get 7 x (1.21+2.52) = 26.1 billion light years. With 8192 intervals we get 41 billion light years. In the limit of very many time intervals we get 42 billion light years. With calculus this whole paragraph reduces to this." from the answer at this site to the 9th question: http://www.astro.ucla .edu/~wright/cosmology_faq.html#DN   **[did you get this?]**

"In an infinite universe, most regions lie beyond our ability to see, even using the most powerful telescopes possible. Although light travels enormously quickly, if an object is sufficiently distant, then the light it emits – will simply not have sufficient time to reach us. Since the universe is about 13.7 billion years old, you might think that anything farther away than 13.7 billion light-years would fall into this category. The reasoning behind this intuition is right on target, but the expansion of space increases the distance to objects whose light has long been traveling and has only just been received, so the maximum distance we can see is actually longer – about 41 billion light-years." Brian Greene, *The Hidden Reality: Parallel Universes and the Deep Laws of the Cosmos*, Alfred A. Knopf, New York, 2011, pages 27 & 28.—

———

— Brian Green goes on to say on page 39 – "As space expands, things dilute and cool, including photons. But unlike particles of matter, photons don't slow down when they cool; being particles of light, they always travel at light speed."

"According to [Princeton University Professor J. Richard] Gott and his collaborators, the radius of the sphere containing all the potentially detectable light sources is approximately 45.7 billion light-years. They performed this calculation by extrapolating back in time to the recombination era, determining the limits of the farthest light sources from that period that we could presently observe, and using the expansion rate of the universe and other cosmological data to determine where those sources are today. Paul Halpern, PhD, *Edge of the Universe: A Voyage to the Cosmic Horizon and Beyond*, John Wiley & Sons, Inc., 2012, page 9.

"One of Hubble's initial 'core' purposes was to determine the rate of expansion of the Universe, known to astronomers as the "Hubble Constant". After eight years of Cepheid observations this work was concluded by finding that the expansion increases with 70 km/second for every 3.26 million light-years you look further out into space."   http://www. spacetelescope.org/science/age_size/

"How can the universe be 95 billion light years across when it has only been in existence for approx 14.3 billion years? If nothing can travel faster than the speed of light shouldn't the universe only measure approx 28.6 billion light-years across?" … "(Since the Big Bang) the universe and space-time itself has been expanding," says astronomer and astrophysicist

Professor Paul Francis from the Australian National University's Mount Stromlo Observatory." … "Since the time of the cosmic microwave background radiation 360,000 years after the Big Bang, the universe and everything in it has expanded away from everything else over 1100 times," he says. "These co-moving co-ordinates as they're called, act like compound interest in finance, dramatically multiplying the size of the observable universe, giving it a diameter of about 93 billion light-years". http://www.abc.net.au/science /articles/2012/02/22/3436134.htm

The right hand of Cosmology does not pay attention to its left. – This inflated edge of space does not correspond to NASA's CMB measurement.

All this means that the Cosmic Microwave Background Radiation could never reach the edge of space some 41 + billion light-year distant compared to the 13.8 billion light years traveled by light photons and cannot be reflected as described by current cosmology and cannot be seen by NASA; consequently the source of the CMB must be the embers or speckles. –The CMB ember radiation has been measured over and over again for the nine years from 2001 – 2011 searching for any errors in this measuring process. No errors found just refinements in detail. ——

——

— we note on page 135 that the CMB ember radiation study with improved detail, had no notation of speckles shifting positions – what one would expect if the CMB was the result of some original Big Bang light being reflected around during the so called expansion of space – noting NASA's nine year observational study. – "The first results [of NASA's CMB] were issued in February 2003, with major updates in 2005, 2007, 2009, 2011, and now this final release." [Page Updated: Monday, 4/08/2013] http://map.gsfc .**nasa**.gov/news

180

A. 27 Forces of Nature – <u>100 % On or 100% Off</u>
Gravity ▪ Electricity ▪ Magnetism ▪ Light Beam ▪

Force of nature – Field
"A field can be considered a type of energy in space, or energy with position. A field is usually visualized as a set of lines surrounding the body, however these lines do not exist, they are strictly a mathematical construct to describe motion. Fields are used in 'electricity, magnetism, gravity and almost all aspects of modern physics.' " http://abyss.uoregon.edu/~js/21st_century_science/lectures/lec04.html

Gravity
"All theoretical and observational studies are completely consistent with the idea that it travels no faster than the speed of light - and no slower." … "If you could turn off gravity, it is mathematically predicted that space and time would also vanish!" http://www.astronomycafe.net/gravity/gravity.html

"As of yet, no technology exists to neutralize the pull of gravity." [In other words, gravity is always 'On.'] http://www.scientificamerican.com/article/fact-or-fiction-anti gravity-chambers-exist/

"Gravity is a force pulling together all matter (which is anything you can physically touch). The more matter, the more gravity" http://www.qrg.northwestern.edu/projects/vss/docs/space-environment/1-what-is-gravity.html

"We can define what it is as a field of influence, because we know how it operates in the universe. And some scientists think that it is made up of particles called gravitons which travel at the speed of light. However, if we are to be honest, we do not know what gravity 'is' in any fundamental way - we only know how it behaves." … "Here is what we do know...

Gravity is a force of attraction that exists between any two masses, any two bodies, any two particles. Gravity is not just the attraction between objects and the Earth. It is an attraction that exists between all objects, everywhere in the universe." http://starchild.gsfc.nasa.gov/docs/StarChild/questions/question30.html

"To begin with, the speed of gravity has not been measured directly in the laboratory" http://math.ucr.edu/home/baez/physics/Relativity/GR/grav_speed.html

Electricity – the flow created is On or Off
"When a wire made of conducting material cuts through a magnetic field, an electrical current is created [in effect turned 'On'] in the wire." http://www.school-for-champions.com/science/electrical_generation.htm#.VPdqFOFwZeE

"Static electricity is an imbalance of electric charges within or on the surface of a material. The charge remains [or is 'off'] until it is able to move away by means of an electric current or electrical discharge [and is 'on']." http://en.wikipedia.org/wiki/Static_electricity

Dangers of static "Static electricity can build up in clouds. [Off] This can cause a huge spark to form between the ground and the cloud. This causes lightning – a flow of charge [On] through the atmosphere." www.bbc.co.uk/schools/gcsebitesize/sci ence/add_edexcel/static_elec/staticrev2.shtml

"A current turns a conductor into an electromagnet. If the current is reversed, the electromagnetic poles will reverse, too. When the electromagnet is placed near a fixed magnet, the two sets of poles repel and attract each other. This produces a force that makes the conductor rotate (spin) at high speed. This turns a shaft, which then drives a machine." http://www.factmonster.com/dk/encyclopedia/electromagnetism.html

182

"In a set of four elegant equations, Maxwell formalized the relationship between electric and magnetic fields. In addition, he showed that a linear magnetic and electric field can be self-reinforcing and must move at a particular velocity, the speed of light." http://abyss.uoregon.edu/~js/21st_century_science /lectures/lec04.html

Magnetism – always 'ON' in each atom – even when two atoms neutralize each other - they are both 'ON' – since the Big Bang
"The magnetic property of a substance arises due to the orbital motion of electrons around the atomic nucleus and intrinsic spin angular momentum. [Always 'On' except –] …   In most materials, these small magnets [atoms] are aligned in such a way, that they cancel each other out." [Off] http://www .buzzle.com/articles/magnetism-vs-gravity.html

"a magnetic field is a change in energy within a volume of space"   https://www.nde-ed.org/EducationResources/Com munityCollege/MagParticle/Physics/MagneticFieldChar.htm

"Charges moving in a magnetic field create an electric field, just as charges moving in an electric field create a magnetic field. This is called electromagnetic induction. Induction provides the basis of everyday technology like transformers on power lines and electric generators."   http://www.sparknotes .com/testprep/books/sat2/physics/chapter16.rhtml

"When you reverse the direction of the current flowing in the wire, the north and south poles are also reversed. When you reverse the current again, the north and south poles reverse again. In fact, each time the current is reversed, the north and south poles will exchange places."   https://www.nde-ed .org/EducationResources/HighSchool/Magnetism/currentflow. htm

"The soft iron inside the coil makes the magnetic field stronger because it becomes a magnet itself when the current is flowing. Soft iron is used because it loses its magnetism as soon as the current stops flowing." http://www.gcsescience .com/pme5.htm

"Inside of a motor, there are essentially four magnets. Two are on opposite sides of the outer casing (the stator), with one that is "pulling" and one that is "pushing." Two other magnets are on opposite sides of the spinning shaft; these switch between one "pulling" and one "pushing" at the same time.
"The idea is that the one of the shaft magnets is set to "push" and the other to "pull," so they are pushed away from the closest stator magnet and pulled towards the next magnet. Just as they get to the halfway point between the two stator magnets, they switch to polarity and are attracted to the next stator magnets. At the exact second that the shaft magnets are closest to the stator magnets, they switch again and are then repelled by the closest magnet and attracted to the next, and continue to rotate. This happens forever as long as the mechanism exists to switch the polarity of the shaft magnets, which we will get to in a little bit." http://www.seaperch .org /electric_motors

"Rotational speeds of up to 1 million rpm – our motor-drive systems are the fastest in the world." [rpm = revolutions per minute] http://www.celeroton.com/en/products.html

"Only certain materials, such as iron, cobalt, nickel, and gadolinium, exhibit strong magnetic effects. These materials are called ferromagnetic. Ferromagnetic materials will respond strongly to magnets and can also be magnetized themselves. "A ferromagnet will lose its magnetism if heated about its Curie temperature. … "The Curie temperature for iron is well above room temperature at 1043 K (770°C)." "Ferromagnets and Electromagnets." *Boundless Physics*. 20

Jan. 2015. https://www.boundless.com/physics/textbooks/boundless-physics-textbook/magnetism-21/magnets-156/ferro magnets-and-electromagnets-551-6041/

Light Beam/Photons

"Photons aren't accelerated *to* the speed of light[1]. Once a photon comes into existence, it is always traveling at the local, constant, speed of light *c*. If you're wondering more about the source of *energy* for photons: remember that a photon is an electromagnetic wave, which is produced by perturbations of an electromagnetic field, e.g. from the acceleration of charged particles." http://physics.stackexchange.com/questions /56818/what-starts-the-movement-of-a-photon

"A photon - as you know - is an electromagnetic (EM) wave . Its energy is described by e = hn [*h* = Plank's constant *n* = frequency] in a quantum of energy. All EM waves travel in the vacuum with the speed of light. A photon is a form of energy and results from transformations of another forms of energy. When the photon appears it behaves like all EM waves." www.newton.dep.anl.gov/askasci /phy00/phy00391.htm

How long must it take to turn 'Space Expansion' On ?

Inside Science News Service, *Shortest Laser Pulse Ever Created:* "American researchers have generated a record-setting laser pulse so short that it makes most everything else seem like an eternity. The pulse lasted just 67 attoseconds, which is about two million billion times faster than the blink of an eye. The previous record, set by European researchers in 2008, was about 20 percent slower. The new record holders say the technique they used can yield even briefer bursts, down to 25 attoseconds, potentially helping physicists see the very motion of electrons around atoms. An attosecond is a billionth of a billionth of a second." http://www.inside science.org/content/shortest-laser-pulse-ever-created/794

185

A. 28 – Inflation, Superstrings, Hidden Dimensions, Parallel Universes, the Multiverse, & Flat Space

Inflation: unable to detect

"…while it is commonly believed that inflation took place in the early Universe, we have been unable so far to detect in experiments on Earth the field responsible for inflation, and so cannot confirm that the proposal for the underlying mechanism is correct." See: Leslie, John, Editor, *Modern Cosmology & Philosophy*, Prometheus Books, 1998, page 276.

inflation is very troublesome

"… the universe, in a moment, expanded at enormous speed. This period of inflation is very troublesome – it requires a new sort of physics to explain it…" See: Seife, Charles, *alpha & omega,* Viking, 2003, page 190.

"Sir Roger Penrose said to me.   …   'I believe the universe has a hyperbolic geometry, but I don't know about the cosmological constant–I don't believe in it.   As for the inflationary universe theory–I am a skeptic.'   …

  Alan Guth counters these arguments with the following: 'The details of exactly how inflation worked are still unclear, but I think that the basic idea of inflation is almost certainly right.' " See: Amir D. Aczel, *God's Equation*, Four Walls Eight Windows, NY, 1999, page 218.

"Despite the prediction inflation [a period of rapid expansion, or inflation, and rapid cooling] as described above is far from an ideal theory. It's too hard to stop the inflationary phase, and the monopole problem has other ways of resurfacing in the physics. Many of the assumptions that go into the model, such as an initial high temperature phase and a single inflating bubble have been questioned and alternative models have been developed." http://www.superstringtheory.com/cosmo/cosmo41.html

"The reason why something like inflation was needed in cosmology was highlighted by discussions of two key problems in the 1970s. The first of these is the horizon problem -- the puzzle that the Universe looks the same on opposite sides of the sky (opposite horizons) even though there has not been time since the Big Bang for light (or anything else) to travel across the Universe and back. So how do the opposite horizons 'know' how to keep in step with each other? The second puzzle is called the flatness problem. This is the puzzle that the spacetime of the Universe is very nearly flat, which means that the Universe sits just on the dividing line between eternal expansion and eventual recollapse." http ://aether.lbl.gov/www/science/inflation-beginners.html

all parts were in contact &   never been in causal contact?
"The first puzzle was the horizon problem. In a hot big bang containing only matter and radiation, the universe flies apart before it can come into equilibrium. Putting that another way, in the simplest picture that one can imagine, all parts of the universe were in contact with all other parts.   But, in the standard big bang model of Lemaître and Hubble, there are region of the universe that are widely separated today, which have never been in causal contact in the past."   Jeremiah P. Ostriker and Simon Mitton, *Heart of Darkness: Unraveling the Mysteries of the Invisible Universe*, Princeton University Press, Princeton, 2013, page 155.

How could anyone say that there was contact and in the same breath that there was no contact? – All the particles were in contact as part of the Lemaître standard model singularity.
See next quote

"When inflation stopped, these regions were so far apart that it looked like they could not have had any casual contact with one another. Yet, in the preinflationary period, they were in casual contact, so it is not a coincidence that the regions look similar." [Inflation only changed the time involved.]   See: Seife, Charles, *alpha & omega,* Viking, N. Y., 2003, pg. 193.

universe [is] an essentially self—created entity
"In the inflationary model of the universe, space originally grew at a very rapid pace. It started out with a tiny initial energy due to one of our vacuum fluctuations. The space that is added as the universe expands has the same properties— curvature, energy density–as the parent space.   The inflationary model—or, rather, the dominant version of that model—sees the universe, the space it occupies, and the energy it contains as an essentially self—created entity." See: Henning Genz, *Nothingness: The science of empty space.* Cambridge, Massachusetts: Perseus books, 1994, page 274.

"… it takes incredible fine-tuning of inflation to give us the world that we live in. The recurring Goldilocks problem reasserts itself. Whether these criticisms show a fundamental flaw in the inflationary paradigm, a flaw from which all inflationary models would suffer, is something that we do not know." Jeremiah P. Ostriker and Simon Mitton, *Heart of Darkness: Unraveling the Mysteries of the Invisible Universe,* Princeton University Press, Princeton, 2013, page 258-259.

"INFLATION has become a cosmological buzzword in the 1990s. No self-respecting theory of the Universe is complete without a reference to inflation -- and at the same time there is now a bewildering variety of different versions of inflation to choose from. Clearly, what's needed is a beginner's guide to inflation, where newcomers to cosmology can find out just what this exciting development is all about. This is it -- new readers start here.
"The Inflation Theory proposes a period of extremely rapid (exponential) expansion of the universe during its first few moments. It was developed around 1980 to explain several

puzzles with the standard Big Bang theory, in which the universe expands relatively gradually throughout its history." http://wmap.gsfc.nasa.gov/universe/bb_cosmo_infl.html

"Surprisingly, the idea of inflation is uncritically accepted by its supporters without there being a definitive physical theory for it. … Therefore an objective assessment is that we do not yet have a theory of inflation that is on as firm a footing as, say, the electro-weak theory. This is the reason why there are many brands of inflation in the marketplace." See: Narlikar, J. & Burbidge, G., *Facts and Speculations in Cosmology,* Cambridge University Press, Cambridge, 2008, page 205

"In the words of Jim Peebles: 'If inflation is wrong, God missed a good trick.'" See: Marcus Chown, *Afterglow of Creation: From the Fireball to the Discovery of Cosmic Ripples,* University Science Books, Sausalito, California, 1996, page 193.

Superstring theory across 11 separate dimensions

"String theory was originated by the Japanese-American physicist Michio Kaku. His theory says that the essential building blocks of all matter as well as all of the physical forces in the universe -- like gravity -- exist on a subquantum level. These building blocks resemble tiny rubber bands -- or strings -- that make up quarks (quantum particles), and in turn electrons, and atoms, and cells and so on. Exactly what kind of matter is created by the strings and how that matter behaves depends on the vibration of these strings. It is in this manner that our entire universe is composed. And according to string theory, this composition takes place across 11 separate dimensions." http://science.howstuffworks.com/science-vs-myth/everyday-myths/parallel-universe3.htm

"Those working on superstrings and other unified theories were not doing physics any more, Glashow [Sheldon Lee, a Nobel Prize winning American theoretical physicist.] contended, because their speculations were so far beyond any possible empirical (based on, concerned with, or verifiable by observation or experience rather than theory or pure logic) test." See: John Hogan, *The End Of Science*: *Facing The Limits Of Knowledge In The Twilight Of The Scientific Age,* Helix Books, Reading, Mass., 1996, page 63.

"Don't be overly impressed by beautiful formalisms. As elegant as string theory appeared, the strings themselves were so tiny that it would still take inconceivably high energies to pry them loose, if they were even there." See: George Johnson, *Strange beauty: Murray Gell-Mann and the revolution in twentieth-century physics*, Alfred A. Knopf, New York, 2000, page 317.

has no experimental evidence in its favor
"Superstring theory has no experimental evidence in its favor. The arguments in favor of superstring theory are really of an aesthetic kind. So far there is no way to test the theory. People just say that it is so beautiful, it has to be true." See: Interview [of Gregory Chaitin] by Hans-Ulrich Obrist The Creative Life: Science vs. Art, October, 2000, page 8. http://www.academia.edu/5847643/Interview_in_Conversations_with_a_Mathematician

Paul Steinhardt: *Einstein: An Edge Symposium*: "My own point of view is that we have to change one or both of these two key components [inflation or string theory] of our understanding of the universe. We either have to dramatically revise them or we have to overhaul them entirely, replace them with something that combines to make a powerful theory that really does explain, in a powerful way, why the universe is the way it is." John Brockman, Editor, *The Universe:*

Leading Scientists Explore the Origin, Mysteries, and Future of the Cosmos, Harper Perenial, New York, 2014, page 248.

no observational or experimental evidence
"There is as yet no observational or experimental evidence for many of the concepts of contemporary theoretical physics, such as super-symmetrical particles, superstrings, the multiverse, the universe as information, the holographic principle or the anthropic cosmological principle." Jim Baggott, *Farewell To Reality: How Modern Physics Has Betrayed the Search for Scientific Truth*, Pegasus Books, New York, 2013, page x.

wishy-washy results – ... – uncalculable by science
"However, two theories in physics, called 'eternal inflation' and 'string theory,' now indicate that the same fundamental principles, from which the laws of nature derive, lead to many different self-consistent universes, with many different properties. ... Such wishy-washy results make theoretical physicists extremely unhappy. Evidently, the fundamental laws of nature do not pin down a single and unique universe. According to the current thinking of many physicists, we are living in one of a vast number of universes. We are living in an accidental universe. We are living in a universe uncalculable by science." See page 7: Alan Lightman, *The Accidental Universe: The World You Thought You Knew*, Pantheon Books, New York, 2013

string theory has — negative results
"So do parallel universes really exist? According to the Many-Worlds theory, we can't truly be certain, since we cannot be aware of them. The string theory has already been tested at least once -- with negative results. Dr. Kaku still believes parallel dimensions do exist, however [source: The Guardian]." http://science.howstuffworks.com/science-vs-myth/everyday-myths/parallel-universe3.htm

"But is our universe unique? The concept of multiple realities — or parallel universes — complicates this answer and challenges what we know about the world and ourselves. One model of potential multiple universes called the Many-Worlds Theory might sound so bizarre and unrealistic that it should be in science fiction movies and not in real life. However, there is no experiment that can irrefutably discredit its validity." http ://www.universetoday.com/113900/parallel-universes-and-the -many-worlds-theory/

desires: "One of his [Prof. Samuel Ting] desires is that the particles recorded by AMS [Alpha Magnetic Spectrometer-2 (AMS) destined for the International Space Station] prove the existence of a parallel universe made up of anti-matter, or particles that are, in electrical charge and magnetic properties, the exact opposite of regular particles. Such a universe has been theorized, but not proven. The discovery of massive amounts of anti-matter could answer fundamental questions about the universe's origin." http://www.nasa.gov/ mission_pages/shuttle/behindscenes/ams_ting.html

hypothetical: "The multiverse (or meta-universe) is the hypothetical set of infinite or finite possible universes (including the universe we consistently experience) that together comprise everything that exists: the entirety of space, time, matter, and energy as well as the physical laws and constants that describe them. The various universes within the multiverse are sometimes called parallel universes or "alternate universes" http://enwikipedia .org/wiki/Multiverse

hypothetical multiverse – anathema to many physicists
"The theory of cosmic inflation, then, supports the scenario in which our universe is just one among many parallel universes in a multiverse. As we will see in later sections, some corroborating evidence for such a scenario also arises from work on dark energy, on superstring theory and on quantum

theory. However, the idea of a hypothetical multiverse, which we can never see or prove, is anathema to many physicists, and many critics still remain." http://www.physicsof theuniverse.com/topics_bigbang_inflation.html

Flat Space
"The simplest version of the inflationary theory, an extension of the Big Bang theory, predicts that the density of the universe is very close to the critical density, and that the geometry of the universe is flat, like a sheet of paper." http:// map.gsfc.nasa.gov/universe/uni_shape.html

A joint Fermilab/SLAC publication© 2014 symmetry, *Our flat universe,* April 07, 2015, " 'The cosmic microwave background in combination with the distribution of galaxies really nails down the flatness,' says Josh Frieman, a physicist at the Fermilab Center for Particle Astrophysics. But, he adds, 'the CMB is the linchpin.' " http://www.symmetrymagazine .org/article/april-2015/our-flat-universe?email_issue=725

there is pretty good evidence that we don't live in ten dimensions:
" … there is pretty good evidence that we don't live in ten dimensions." Steven Weinberg, *Lake Views: This World and the Universe*, The Belknap Press of Harvard University Press, Cambridge, 2009, page 30.

A. 29 – Quasars

"A Quasar is an enormously bright object at the edge of our universe with the appearance of a star when viewed through a telescope. It emits massive amounts of energy, more energy than 100 normal galaxies combined. The name comes from a shortening of quasi-stellar radio source (QSR). Current theories hold that quasars are one type of active galactic nuclei (AGN). Many astronomers believe supermassive black holes may lie at the center of these galaxies and power their explosive energy output. In one second, a typical quasar releases enough energy to satisfy the electrical energy needs of Earth for the next billion years. It is thought by many astronomers that quasars are the most distant objects yet detected in the Universe. With the massive amounts of energy a quasar emits, it can be a trillion times brighter than our own sun." http://space.about.com/od/deepspace/a/quasarinfo.htm

"Many astronomers believe that quasars are the most distant objects yet detected in the universe. Quasars give off enormous amounts of energy - they can be a trillion times brighter than the Sun! Quasars are believed to produce their energy from massive black holes in the center of the galaxies in which the quasars are located. Because quasars are so bright, they drown out the light from all the other stars in the same galaxy." http://starchild.gsfc.nasa.gov/docs/StarChild/universe_level2/quasars.html

"The quasar that has just been found, named ULAS J1120+0641, is seen as it was only 770 million years after the Big Bang (redshift 7.1). It took 12.9 billion years for its light to reach us.
"Although more distant objects have been confirmed (such as a gamma-ray burst at redshift 8.2 and a galaxy at redshift 8.6), the newly discovered quasar is hundreds of times brighter than

these." This research was presented in a paper to appear in the journal *Nature* on 30 June 2011.   http://www.sciencedaily com/releases/2011/06/110629132527.htm

Failed to Find

"In a recent survey with the Hubble Space Telescope Bahcall *et al.* (1994) failed to find the so-called *'host galaxy'* surrounding many quasars. The standard cosmological theory of quasars predicted that the nebulosity should have easily been detected with the resolution and dynamic range of the Hubble Space Telescope. This is a completely unexpected discovery and cosmologists are dumbfounded:" http://laser stars.org/news/NakedQSO.html

"Hubble astronomers have looked at one of the most distant and brightest quasars in the universe and are surprised by what they did not see: the underlying host galaxy of stars feeding the quasar."   http://www.nasa.gov/mission_pages/webb /news/dusty-quasar.html

A. 30 Galaxies need time to form, Black Holes, & Hubble's Youngest Galaxies are at least 13 billion years old.

The biggest problem with scientific thought about how galaxies are formed is due to the assumed concept that the Big Bang took place only some 13+ billion years ago. This book's analysis expands our understanding that the Big Bang took place some 28 billion years ago providing gravity all the time necessary to form galaxies.

"One of the criticisms that Lerner [Eric Lerner of Lawrence Plasma Laboratory in New Jersey] has given time and again is that there is not enough time in the big bang age of the universe (approximately 15 billion years) for the observed structures to have formed." See: Barry Parker, *The Vindication of the Big Bang: breakthroughs and barriers*, Plenum Press, New York, 1993, page 310.

"Jeremaih Ostriker, a respected theorist and the chairman of Princeton University's Department of Astrophysical Sciences told me at the time: 'No existing theory can explain the structure of the universe. There is some missing ingredient, some crucial fact that we haven't uncovered, and that's becoming more apparent all the time.' " See: Michael Lemonick, *The Light at the Edge of the Universe: Leading Cosmologists on the Brink of a Scientific Revolution.* New York: Villard Books, 1993, page 17.

"When Gamow first proposed what later would become known as the hot big bang concept, the difficulties of understanding galaxy formation in a rapidly expanding universe stood in the way of wider acceptance of the big bang model. The more one thinks about this problem, the more puzzling it becomes. ... Whatever the answer to this question (and the jury is still out), it seems that conditions did favor the creation of galaxies." Pages 138 & 139 ... & on to page 262

"The origins of both of the major components [dark matter & 'dark energy'] of the universe and the seed fluctuations that grew to become the things [galaxies] we see about us continue to baffle us." Jeremiah P. Ostriker and Simon Mitton, *Heart of Darkness: Unraveling the Mysteries of the Invisible Universe*, Princeton University Press, Princeton, 2013.

"In 1963, Hoyle and his two collaborators Margaret and Geoffrey Burbridge summarized the situation as follows: 'Undoubtedly, the greatest shortcoming of all cosmological theories lies in their failure to provide a working model of the formation of galaxies.' " See: Helge Kragh, *Cosmology and Controversy,* Princeton University Press, 1996, page 295.

"Whatever reality you subscribe to, neither the big bang nor the cosmological principle explains the existence of galaxies." Jeff Kanipe, *Chasing Hubble's shadows: the search for galaxies at the edge of time,* Hill and Wang, N.Y., 2006, p. 85.

"It was a big surprise that these galaxies are out there and they're full of old stars." Says [Roberto] Abraham. "But it's not clear that you can't explain them within a hierarchical galaxy-formation picture, provided you come up with a means of making star formation occur surprisingly early in the universe and then just shutting it off. You have a big episode where you make the galaxy at redshift 3, 4, or 5, but it only forms all its stars for about 200 million years and then just shuts off. At the moment we have to come up with a mechanism for doing that, and nobody's come up with that so far." J. Kanipe, *Chasing Hubble's Shadows,* 2006, page 143.

"Today's universe is lumpy, not smooth, with matter clumped into galaxies and clusters of galaxies. Scientist concluded that there had not been enough time for the evolution of so much large-scale structure by normal gravitational processes unless there had been some kind of irregularity, or 'anisotropy,' in the earlier stage of the universe." John Noble Wilford, *Cosmic Dispatches: The New York Times Reports on Astronomy and Cosmology.* New York: W. W. Norton & Co., 2001, page 211. [anisotropy - directionally dependent]

"Eventually we might hope that this whole process of galaxy formation and evolution will be understood as a natural consequence of the initial density fluctuations, the nature of dark matter in the universe, and the cosmological parameter." See: Michael Rowan-Robinson, *The Nine Numbers of the Cosmos,* Oxford University Press, 1999, page 147.

"A cloud of gas will tend to contract under its own gravity and then fragment into smaller clouds … Unfortunately, the growth of galaxies by this mechanism would take much longer than the age of the universe [estimated in the 20th century to be ~14-15 billion years]" See: John Leslie, Editor, *Modern Cosmology & Philosophy*, Prometheus Books, 1998, page 239.

"But even some of the proponents of cold dark matter— including [Jim] Peeples, who invented the idea—have begun to worry just a little. 'The cold dark matter model of galaxy formation is in deep trouble,' he says." See: Marcus Chown, *Afterglow of Creation: From the Fireball to the Discovery of Cosmic Ripples.* Sausalito, California: University Science Books, 1996, page 192.

"Indeed, almost all of our attempts to decipher the geometry of the universe and its fate have been frustrated by our lack of understanding of how galaxies evolve." See: Joseph Silk, *A Short History of the Universe.* New York: scientific American Library, 1994, page 214.

Black Holes
"Though much more analysis remains, an initial look at Hubble evidence favors the idea that titanic black holes did

not precede a galaxy's birth but instead co-evolved with the galaxy by trapping a surprisingly exact percentage of the mass of the central hub of stars and gas in a galaxy." News Release Number: STScI 2000-22. http://hubblesite.org/newscenter /archive/releases/2000/22/

"Black holes are objects so dense, and with so much mass, that even light cannot escape their gravity. ...
"Hubble has also proved that super massive black holes are most likely present at the centres of most, if not all, large galaxies. This has important implications for the theories of galaxy formation and evolution. ...
"There must be some mechanism that links the formation of the galaxy to that of its black hole and vice versa. This has profound implications for theories of galaxy formation and evolution and is an ongoing area of research in astronomy.
"One big question which remains is why most galaxies in our cosmic neighbourhood, including the Milky Way, appear to have a dormant black hole which is not funnelling in large amounts of matter at present." www.spacetelescope.org /science/black_holes/

"Thus it's now believed that black holes are not only common throughout the Cosmos but they play a fundamental role in the formation and evolution of the Universe we inhabit today." http://www.cosmotography.com/images/supermassive _blackholes_drive_galaxy_evolution_2.html

Hubble sees tiny galaxies
"Scientists analyzed the image statistically and found that the HDF had seen back to the very young Universe where the bulk of the galaxies had not, as yet, had time to form stars. Or, as the popular press dramatically reported, "Hubble sees back to Big Bang". These very remote galaxies also seemed to be smaller and more irregular than those nearer to us. This was

taken as a clear indication that galaxies form by gravitational coalescence of smaller parts." http://www.spacetelescope .org/science/deep_fields/

"The deeper Hubble looks into space, the farther back in time it looks, because light takes billions of years to cross the observable universe. This makes Hubble a powerful "time machine" that allows astronomers to see galaxies as they were 13 billion years ago, just 600 million to 800 million years after the Big Bang. ... "The data from Hubble's new infrared camera, the Wide Field Camera 3 (WFC3), on the Ultra Deep Field (taken in August 2009) have been analyzed by no less than five international teams of astronomers. ... The existence of these newly found galaxies pushes back the time when galaxies began to form to before 500-600 million years after the Big Bang. ... These galaxies are as small as 1/20th the Milky Way's diameter," reports Pascal Oesch of the Swiss Federal Institute of Technology in Zurich. ... The HUDF09 team also combined the new Hubble data with observations from NASA's Spitzer Space Telescope to estimate the ages and masses of these primordial galaxies. "The masses are just 1 percent of those of the Milky Way," explains team member Ivo Labbe of the Carnegie Institute of Washington, leader of two papers on the data from the combined NASA Great Observatories. He further noted that "to our surprise, the results show that these galaxies at 700 million years after the Big Bang must have started forming stars hundreds of millions of years earlier, pushing back the time of the earliest star formation in the universe." ... The clear detection of galaxies between $z=7$ and $z=8.5$ corresponds to "look-back times" of approximately 12.9 billion years to 13.1 billion years ago." News Release Number: STScI-2010-02 Jan. 5, 2010 Space Telescope Science Institute, Baltimore, Md. http:// hubblesite.org/newscenter/archive/releases/2010 /02 /full/

Hubble's Youngest Galaxies   What Hubble sees are small irregular galaxies older than the CMB Embers:

"The most distant galaxies look strange – smaller, irregular, lacking clearly defined shapes."   http://hubblesite.org /hubble_ discoveries /breakthroughs /cosmology

The Assumption

"The universe is 13.7 billion years old, and the XDF [eXtreme Deep Field] reveals galaxies that span back 13.2 billion years in time. Most of the galaxies in the XDF are seen when they were young, small, and growing, often violently as they collided and merged together. The early universe was a time of dramatic birth for galaxies containing brilliant blue stars extraordinarily brighter than our sun. The light from those past events is just arriving at Earth now, and so the XDF is a 'time tunnel into the distant past.' [**Here is the Assumption -**] The youngest galaxy found in the XDF existed just 450 million years after the universe's birth in the big bang." http://hyper physics.phy-astr.gsu.edu/hbase /astro/xdeepfield.html

What Hubble sees are galaxies that are 13.2 billion light years away from Earth. Traveling at near the speed of light the Big Bang embers took at least 13.2+ billion years to form these youngest galaxies – not 450 million years. What this all means is that these youngest galaxies seen by Hubble are at least 13 billion years old.

A 31 Thermonuclear ElectroMagnetic Pulse [EMP], Chaos, Turbulence, The Butterfly Effect & No Perfect Circle

"Although nuclear EMP was known since the very first days of nuclear weapons testing (and often caused problems in the local area-- especially with monitoring equipment), the magnitude of the effects of high-altitude nuclear EMP were not known until a 1962 test of a thermonuclear weapon in space called the Starfish Prime test. The Starfish Prime test knocked out some of the electrical and electronic components in Hawaii, particularly in Honolulu, which was 897 miles (1445 kilometers) away from the nuclear explosion." www.fut urescience.com/emp.html

"An electromagnetic pulse is a burst of electromagnetic radiation. Nuclear explosions create a characteristic pulse of electromagnetic radiation called a Nuclear EMP or NEMP. …
"The resulting rapidly changing electric fields and magnetic fields may couple with electrical/electronic systems to produce damaging current and voltage surges." http://en.wikipedia. org/wiki/Nuclear_electromagnetic_pulse

"The pulse can easily span continent-sized areas, and this radiation can affect systems on land, sea, and air. The first recorded EMP incident accompanied a high-altitude nuclear test over the South Pacific and resulted in power system failures as far away as Hawaii. A large device detonated at 400–500 km [250 to 300 miles]over Kansas would affect all of CONUS." [CONUS – the 48 contiguous states]." http://fas.org /nuke /intro/nuke/emp.htm

"EMP from high-yield nuclear detonations will subject electrical grids to voltage surges far exceeding those caused by lightning." http://www.nucleardarkness.org/nuclear/effects ofnuclearweapons/

"Between August and September 1958, the US Navy exploded three fission type nuclear bombs 480 km [300 miles] above the South Atlantic Ocean, in the part of the lower Van Allen Belt closest to the earth's surface. In addition, two hydrogen bombs were detonated 160 km [100 miles] over Johnston Island in the Pacific. The military called this "the biggest scientific experiment ever undertaken." It was designed by the US Department of Defense and the US Atomic Energy Commission, under the code name Project Argus. The purpose appears to be to assess the impact of high altitude nuclear explosions on radio transmission and radar operations because of the electromagnetic pulse (EMP), and to increase understanding of the geomagnetic field and the behavior of the charged particles in it." http://www.theforbiddenknowledge com/hardtruth/haarp_mind_weather_control.htm

"High Power Electromagnetic Pulse generation techniques and High Power Microwave technology have matured to the point where practical E-bombs (Electromagnetic bombs) are becoming technically feasible, with new applications in both Strategic and Tactical Information Warfare. The development of conventional E-bomb devices allows their use in non-nuclear confrontations. This paper discusses aspects of the technology base, weapon delivery techniques and proposes a doctrinal foundation for the use of such devices in warhead and bomb applications." http://www.ausairpower.net/ASPC-E-Bomb-Mirror.html

"The Soviet Union had first suffered electromagnetic pulse (EMP) damage to electronic blast instruments in their 1949 test. Their practical understanding of EMP damage eventually led them, on Monday 22 October 1962, to detonate a 300 kt missile-carried thermonuclear warhead at an altitude of 300 km (USSR test 184)." http://ed-thelen.org/EMP-ElectroMag neticPulse.html   [300 kilometers = 186.411 miles]

Turbulent Flows: General Properties

" … you would expect anything as chaotic as the Big Bang to produce a highly turbulent cloud of expanding gas. There would be swirls and eddies in the plasma, and these would correspond to ready-made concentrations of mass to start the accumulation process. This idea was quite popular for a while, but numerous calculations showed that even starting off with normal turbulence in the plasma, the gravitational attraction was still too slow for galaxies to form." James S. Trefil, *The Moment of Creation: Big Bang Physics From Before the First Millisecond to the Present Universe*, Charles Scribner's Sons, New York, 1983, pages 43

"Turbulence is ubiquitous in natural fluids: atmosphere, ocean, lakes, rivers, Earth's interior, planetary atmospheres and their convective interiors, stars, and space gases (neutral and ionized). … the generic behavior of the entanglement of neighboring material parcels; this causes chaotic evolution, transport, and mixing … Turbulent flows manifest a complexity that has thus far exceeded scientists' abilities to measure, theorize, or simulate comprehensively." http:// web.atmos.ucla.edu/~jcm/turbulence_course_notes/turbulent _flows.pdf

"One of them, to take a problem within physics, is to understand the flow of a fluid when it becomes turbulent. This problem has faced us for one hundred years; it still defies solution and may go on resisting solution well after the success of the final theory of elementary particles, because we already understand all we need to know about the fundamental principles governing fluids." Steven Weinberg, *Lake Views: This World and the Universe*, The Belknap Press of Harvard University Press, Cambridge, 2009, page 33.

"Flows of fluids—liquids and gases—generally become turbulent once they start flowing fast enough. When they flow slowly, all of the fluid moves in parallel, rather like ranks of marching soldiers. But as the speed increases, the ranks break up. You could say that the "soldiers"—little parcels of fluid—begin to bump into one another or move sideways, and so swirls and eddies begin to form."   …

"When the flow is turbulent, this interdependence is extreme and the flow becomes chaotic, in the technical sense that the smallest disturbances at one time can lead to completely different patterns of behavior at a later moment."   http://nautil.Us/issue/15/turbulence/the-scientific-problem-that-must-be-experienced

"Turbulence is composed of eddies: patches of zigzagging, often swirling fluid, moving randomly around and about the overall direction of motion. Technically, the chaotic state of fluid motion arises when the speed of the fluid exceeds a specific threshold, below which viscous forces damp out the chaotic behavior."   http://www.cascadetechnologies.com/turbulent-motions/

Chaos Theory   &   The Butterfly Effect
"Chaos Theory deals with nonlinear things that are effectively impossible to predict or control, like turbulence, weather, the stock market, our brain states, and so on. These phenomena are often described by fractal mathematics, which captures the infinite complexity of nature. Many natural objects exhibit fractal properties, including landscapes, clouds, trees, organs, rivers etc, and many of the systems in which we live exhibit complex, chaotic behavior."
"The Butterfly effect   …   small changes in the initial conditions lead to drastic changes in the results."
http://fractalfoundation.org /resources/what-is-chaos-theory/

No Perfect Circle
[[[A perfect circle, built on the unending number - π, as far as physics knows, exists only in a mathematicians mind.]]]

206

Rebel — a person who rises in opposition. Cosmologist cannot be a rebel without putting his career at risk.
"Einstein was also something of a rebel, ready and willing to challenge prevailing opinions. He was at this time completely unknown to the academic establishment, and could therefore publish his outrageous ideas without fear of putting his (non-existent) academic reputation at risk." Jim Baggott, *Farewell To Reality: How Modern Physics Has Betrayed the Search for Scientific Truth*, Pegasus Books, New York, 2013, page 85.

"no way to describe scientifically the origin ..."
"There is no way to describe scientifically the origin of the universe without treading upon territory held for millennia to be sacred. Beliefs about the origin of the universe are at the root of our consciousness as human beings. This is a place where science, willingly or unwillingly, encounters concerns traditionally associated with a spiritual dimension." Joel R. Primack, Professor of Physics, University of California, Santa Cruz. [UCSC] http://physics.ucsc.edu/cosmo/primack_abrams /COSMO.HTM

[radical or rebel by definition cannot be conservative!!!]
"I [Frank J. Tippler, a professor of mathematical physics at Tulane University] also realized that to be a radical in science, one needed to be a conservative;   that is, revolutions in physics are accomplished not by deliberately trying to overthrow known physics but by thinking deeply about the full implications of the laws of physics that we believe we know." See: Edited by John Brockman, *My Einstein: Essays by Twenty-four of the World's Leading Thinkers on the Man, His Work, and His Legacy.* Pantheon Books, 2006, page 76.

Radical: adjective:
"(especially of change or action)

"a radical overhaul of the existing regulatory framework" …
"innovative or progressive"
https://www.google.com/#hl=en&q=definition+-+radical

Conservative: noun "a person who is averse to change and holds to traditional values and attitudes"
https://www.google.com/#hl=en&q=definition+-+conservative

Cannot change pre-conditioning Conservative to Radical
"changing our beliefs is not an easy thing to do. Most people find changing one small belief extremely difficult, let alone a whole range of self-supporting beliefs based on negative pre-conditioning."   http://www.mind-sets.com/html/mindset /how_to_change_your_mindset.htm

cards are stacked against the revolutionary
"More than at any time in the history of science, the cards are stacked against the revolutionary. Such people are simply not tolerated in the research universities. Little wonder, then that even when the science clearly calls for one, we can't seem to pull off a revolution." See: Lee Smolin, *The Trouble with Physics: the rise of string theory, the fall of a science, and what comes next,* A Mariner Book, Boston, 2006, page 348.

"To be doing astrophysics you are going to want to get a tenured faculty position at a premiere or 1st division university. You could alternatively get a position at a premiere non-university research institution (NASA, Perimeter Institute, etc.) - this is also just as hard." How to get a job in astronomy / cosmology / astrophysics (Koohii Lounge)   http://forum. koohii.com/viewtopic.php?id=11166

Theoretical physicist is KEY word for conservative
"We are looking for a theoretical physicist in the areas of collider physics, QCD, beyond the Standard Model, quantum field theory and string theory, heavy flavor physics and astroparticle physics/cosmology. … Qualification required: PHD/doctorate in the field of physics, or equivalent.

The experience required for this post is: Thorough knowledge of theoretical physics. [CERN is the European Organization for Nuclear Research.]   http://jobs.web.**cern**.ch/job/11019

"...I accumulated various numbers showing the way hiring in particle theory at leading institutions in the US.   ... there's a much larger number of talented and accomplished candidates than there are jobs, and departments are playing it safe, offering the few jobs available only to people working in a small number of areas that are conventionally agreed to be "hot". As always, if you're working on some idea that's not in the narrow mainstream, there's no chance you'll get hired into a permanent position at a US institution." www.math.columbia .edu/~woit/wordpress/?p=4701

"Additional experience and training in a postdoctoral research appointment, although not required, is important for physicists and astronomers aspiring to permanent positions in basic research in universities and government laboratories. Many physics and astronomy Ph.D. holders ultimately teach at the college or university level."   http://www.answers.com/ Q/What_is_the_educational_requirement_for_a_cosmology

Science Is Not About Certainty: The separation of science and the humanities is relatively new—and detrimental to both by Carlo Rovelli: "Restricting our vision of reality today to just the core content of science or the core content of the humanities is being blind to the complexity of reality, which we can grasp from a number of points of view." www.newrepublic.com/article/118655/theoretical-phyisicist-explains-why-science-not-about-certainty

"Astronomers spend most of their time analyzing data with computers. They also are often teachers at colleges and universities."   http://curious.astro.cornell.edu/people-and-astronomy/careers-in-astronomy

"Cosmologists do not usually work within the framework of alternative cosmologies because they feel that these are not at present as competitive as the standard model. Certainly, they are not so developed, and they are not so developed because cosmologists do not work on them. It is a vicious circle. The fact that most cosmologists do not pay them any attention and only dedicate their research time to the standard model is to a great extent due to a sociological phenomenon (the "snowball effect" or "groupthink")."   http ://arxiv.org/abs /1311.6324

"Why, after all, should the student of physics, for example, read the works of Newton, Faraday, Einstein, or Schrödinger, when everything he needs to know about these works is recapitulated in a far briefer, more precise, and more systematic form in a number or up-to-date textbooks?" Thomas S. Kuhn, *The Structure of Scientific Revolutions*, Third Edition, University of Chicago Press, 1996, page 165.

"According to [Thomas S.] Kuhn the scientific establishment is defined by its members' belief in the set of prevailing theories, which together form a world-view [such as the current concept that our universe was created out of nothing], or *paradigm*.   ...   "Should any observation seem to violate the relevant paradigm, its holders are simply blind to the violation.   When confronted with evidence of it, they are obliged to regard it as an 'anomaly', an experimental error, a fraud – anything at all that will allow them to hold the paradigm valid.   ...   "Where applicable, they may well use methods that are scientific in the Popperian sense, but they never discover anything fundamental because they never question anything fundamental." **[such as the assumption that our Universe was created from nothing.] [science needs a rebel!]**   David Deutsch, *The Fabric of Reality: The Science of Parallel Universes – and Its Implications*, Allen Lane, The Penguin Press, New York, 1997, pages 321, 322.

Bibliography

Amir D. Aczel, *God's Equation: Einstein, Relativity, and the Expanding Universe*, Four Walls Eight Windows, New York, 1999.

Peter J. Aubusson, Allan G. Harrison, Stephen M. Ritchie, Editors, *Metaphor and Analogy in Science Education*, Springer; The Netherlands, 2006.

Jim Baggott, *Farewell To Reality: How Modern Physics Has Betrayed the Search for Scientific Truth*, Pegasus Books, New York, 2013.

Lincoln Barnett, *The Universe and Dr. Einstein: with a forward by Albert Einstein.* Mattituck, American Reprint Company, N.Y., 1950.

John D. Barrow, *The Origin of the Universe,* Basic Books, New York, 1994.

David Berlinski, *Was There a Big Bang?* New York: Commentary Magazine, February 1998.

John Brockman, Editor, *My Einstein: Essays by Twenty-four of the World's Leading Thinkers on the Man, His Work, and His Legacy,* New York: Pantheon Books, 2006.

————, *The Universe: Leading Scientists Explore the Origin, Mysteries, and Future of the Cosmos*, Harper Perenial, New York, 2014.

Gregory J. Chaitin, *Conversations with a Mathematician: Math, Art, Science and the limits of Reason.* Great Britain: Springer, 2002.

Marcus Chown, *Afterglow of Creation: From the Fireball to the Discovery of Cosmic Ripples,* University Science Books, California, 1996.

K. C. Cole, *The Hole in the Universe: How Scientists Peered over the Edge of Emptiness and Found Everything.* New York: Harcourt, Inc., 2001.

Ken Croswell, *The Universe at Midnight: Observations Illuminating the Cosmos.* New York: The Free Press, 2001.

Christopher De Pree, PhD, *THE COSMOS: Idiot's Guides as Easy as It Gets!,* Alpha, New York, 2014.

David Deutsch, *The Fabric of Reality: The Science of Parallel Universes — and Its Implications*, Allen Lane, Penguin Press, New York, 1997.

—————, *The Beginning of Infinity: Explanations that Transform the World.* New York: Viking, 2011.

Terence Dickinson, *The Universe and Beyond.* Buffalo, New York: A Firefly Book, 1999.

Albert Einstein; G B Jeffery; Wilfrid Perrett, *Sidelights on relativity. I. Ether and relativity. II. Geometry and experience,* London, 1922

Richard Feynman, *Six Easy Pieces: Essentials of Physics Explained by Its Most Brilliant Teacher.* Helix Books, Massachusetts, 1963.

—————, *The Meaning of It All: Thoughts of a Citizen Scientist,* Helix Books/Perseus Books, Massachusetts, 1999.

—————, *The Character of Physical Law*, M.I.T. Press, Massachusetts: 1965.

Richard Phillips Feynman and Jeffrey Robbins, *The Pleasure of Finding Things Out: The Best Short Works of Richard P. Feynman,* by, Basic books, New York, N. Y. 2005.

Richard P. Feynman and Christopher Sykes, *No ordinary genius: the illustrated Richard Feynman*, Norton, New York, 1994.

Pedro G. Ferreira, *The Perfect Theory: A Century of Geniuses and the Battle over General Relativity*, Houghton Mifflin Harcourt, Boston, 2014.

Karen C. Fox, *The Big Bang Theory: What it is, Where it Came From, and Why it works.* New York: John Wiley & Sons, Inc. 2002.

Harald Fritzsch, *An Equation That Changed the World: Newton, Einstein, and the Theory of Relativity,* The University of Chicago Press, 1994.

Maurizio Gasperini, *The Universe Before the Big Bang: Cosmology and String Theory*, Springer, Berlin, 2008.

Henning Genz, *Nothingness: The science of empty space.* Perseus books, Massachusetts, 1994.

James Gleick, *Genius: The Life and Science of Richard Feynman.* New York: Pantheon Books, 1992.

Martin Gorst, *Measuring Eternity: The Search for the Beginning of Time,* Broadway book, New York, 2001.

Brian Greene, *The Hidden Reality: Parallel Universes and the Deep Laws of the Cosmos*, Alfred A. Knopf, N. Y., 2011.

John and Mary Gribbin, *Richard Feynman: A Life in Science.* New York: A Dutton Book, 1998.

Alan Guth, in the Fall, 1997 SLAC National Accelerator Laboratory publication *the Beam Line* article titled: *Was the Cosmic Inflation the 'Bang of the Big Bang?*

————, *The Inflationary Universe,* Basic Books, NY, 1998.

Paul Halpern, PhD, *Edge of the Universe: A Voyage to the Cosmic Horizon and Beyond*, John Wiley & Sons, Inc., 2012.

Richard Hammond, PhD, *The Unknown Universe: The Origin of the Universe, Quantum Gravity, Wormholes, and Other Things Science Still Can't Explain*, New Page Books, A division of the Career Press, Inc., NJ. 2008.

Stephen Hawking, *The Universe in a Nutshell*, Bantam Books, New York, 2001.

————, *A Stubbornly Persistent Illusion*, Running Press, Philadelphia, 2007.

John Hogan, *The End Of Science: Facing The Limits Of Knowledge In The Twilight Of The Scientific Age,* Helix Books, Reading, Mass., 1996.

Edwin Hubble, *The Realm of the Nebulae*, Dover Publications Inc., New York, 1958 by Dover [originally by Yale 1936].

Chris Impey, *How It Began: A Time-Traveler's Guide To The Universe*, W. W. Norton & Company, New York, 2012.

Robert Jastrow, *God and the Astronomers.* New York: Warner Books, 1978.

George Johnson, *Strange beauty: Murray Gell-Mann and the revolution in twentieth-century physics*, Alfred A. Knopf, New York, 2000.

Jeff Kanipe, *Chasing Hubble's Shadows: the search for galaxies at the edge of time.* NY: Hill and Wang, 2006.

Morris Kline, *Mathematics: The Loss of Certainty.* Oxford: Oxford University Press, 1980.

Robert P. Kirshner, *The Extravagant Universe: exploding stars, dark energy, and the accelerating cosmos*, Princeton University Press, Princeton, 2002.

Helge Kragh, *Cosmology and Controversy: The Historical Development of Two Theories of the Universe.* Princeton, N.J.: Princeton University Press, 1996.

Lawrence M. Krauss, *A Universe From Nothing: Why There Is Something Rather than Nothing.* New York: Free Press, 2012.

—————————, *Quantum Man; Richard Feynman's Life in Science,* New York, W. W. Norton & Co., 2012.

Thomas S. Kuhn, *The Structure of Scientific Revolutions*, Third Edition, University of Chicago Press, Chicago, 1996.

Abbé Georges-Henri Lemaître, *The Beginning of the World from the Point of View of Quantum Theory*, Nature, March 21, 1931.

Michael Lemonick, *The Light at the Edge of the Universe: Leading Cosmologists on the Brink of a Scientific Revolution,* Villard Books, New York, 1993.

Eric J. Lerner, *The Big Bang Never Happened: a Startling Refutation of the Dominant Theory of the Origin of the Universe.* Time Books, 1990.

John Leslie, Editor, *Modern Cosmology & Philosophy*, Prometheus Books, Amherst, N.Y., 1998.

Alan Lightman, *The Accidental Universe: The World You Thought You Knew*, Pantheon Books, New York, 2013.

David Lindley, *The End of Physics: The Myth of a Unified Theory.* New York: Basic Books, 1993.

J. P. McEvoy, *A Brief History of the Universe*, Robinson Publishing, London, 2010.

Arthur I. Miller, *Albert Einstein's Special Theory of Relativity: Emergence (1905) and Early Interpretation (1905—1911)*, Addison-Wesley Publishing Company, Inc. 1981.

Narlikar, J. & Burbidge, G., *Facts and Speculations in Cosmology,* Cambridge University Press, Cambridge, 2008.

Sharon L. Nichols, T. L. Good, *America's Teenagers--Myths and Realities*: *Media Images, Schooling, and the Social Costs of Careless Indifference,* Routledge, N.J., 2004.

Jeremiah P. Ostriker and Simon Mitton, *Heart of Darkness: Unraveling the Mysteries of the Invisible Universe*, Princeton University Press, Princeton, 2013.

Barry Parker, *The Vindication of the Big Bang: breakthroughs and barriers*, Plenum Press, New York, 1993.

P.J.E. Peebles, *Principles of Physical Cosmology*, Princeton University Press, Princeton, 1993.

Lisa Randall, *Knocking on Heaven's Door, How Physics and Scientific Thinking Illuminate the Universe and the Modern World*, HarperCollins, New York, 2011.

Michael Rowan-Robinson, *Ripples in the Cosmos,* Oxford: Oxford University Press, 1998.

―――――――――――, *The Nine Numbers of the Cosmos,* Oxford, Oxford University Press, 1999.

Charles Seife, *Zero: the biography of a dangerous idea,* Viking, New York, 2000.

―――――――, *alpha & omega,* Viking, New York, 2003.

Harvey Siegel, Editor, *Reason and Education: Essays in Honor of Israel Scheffler*, Kluwer Academic Publisher, The Netherlands, 1997.

Joseph Silk, *On the Shores of the Unknown: A Short History of the Universe*, Cambridge University Press, New York, 2005.

Joseph Silk, *The Big Bang*, W. H. Freeman and Company, NY, 1989.

—————, *Horizons of Cosmology: Exploring Worlds Seen and Unseen*, Templeton Press, 2009.

Lee Smolin, *The Trouble with Physics: the rise of string theory, the fall of a science, and what comes next*, A Mariner Book, Boston, 2006.

Max Tegmark, *Our Mathematical Universe*, New York, Alfred A. Knopf, 2014.

James S. Trefil, *The Moment of Creation: Big Bang Physics From Before the First Millisecond to the Present Universe*, Charles Scribner's Sons, New York, 1983.

È. A Tropp, Viktor ÍAkovlevich Frenkel'; A D Chernin, *"Alexander A Friedmann: The Man Who Made the Universe Expand"* Cambridge: Cambridge University Press, 1993.

Jenny Volvovski, Julia Rothman, and Matt Lamothe, *The Where, The Why, and The How: 75 artists Illustrate Wondrous Mysteries of Science*, Chronicle Books, San Francisco, 2012.

Steven Weinberg, *Lake Views: This World and the Universe*, The Belknap Press of Harvard University Press, Cambridge, 2009.

—————, *Dreams of a Final Theory.* New York: Pantheon Books, 1994.

—————, *The First Three Minutes: A Modern View of the Origin of the Universe,* New York: Basic Books, 1977.

David A. Weintraub, *How Old is the Universe?,* Princeton University Press, Princeton, 2011.

John Noble Wilford, *Cosmic Dispatches: The New York Times Reports on Astronomy and Cosmology.* New York: W. W. Norton & Co., 2001.

Anthony Zee, *Einstein's Universe: Gravity at work and Play.* New York: Oxford University Press, 1989.

"There is no way to describe scientifically the origin of the universe without treading upon territory held for millennia to be sacred." Joel R. Primack, Physics Prof. [ref-p207]   "Or without treading upon assumptions held sacrosanct by cosmologists." Charles Sven

Only a rebel can present to the world a reality view [this book describing How, Where & When the Big Bang Banged] that is in direct opposition to those old mathematical assumptions held by cosmologists that say our Universe started from nothing.

Charles is that rebel

Notes on front and back cover images

The front cover of this book presents the NASA <u>full sky</u> version normally reported of the Cosmic Microwave Background Radiation. [CMB]

In trying to show all; this CMB is distorted because that's what happens when one puts a 3D observation on a 2D flat surface. [Front cover is a Mollweide Projection of the CMB] This distortion can only be corrected by projected that full sky image on a globe or celestial sphere, this back cover.

[A copy of a celestial sphere, a projection of the coordinates of the night sky as seen from Earth, can be seen on page 42.]

This back cover presents one half of the CMB sky set on a celestial sphere rarely seen in NASA reports but presented here for the purpose of presenting the best 3D depiction [or least distorted image] of our Universe with a center wedge cut out to display NASA's observations and measurements of the CMB surrounding centrally located Earth [see page 45.]

Triangulating [see page 44] the NASA data of celestial coordinates and distances places our Earth virtually in the center of our observable Universe and is totally supported by our isotropic view of galaxies and quasars [see p. 66, & 67.]

This back cover image fully supports this book's analysis found on page 73 that concludes that:

We Earthlings are located right next to the Big Bang Epicenter of our observable Universe.

www.ingramcontent.com/pod-product-compliance
Lightning Source LLC
Chambersburg PA
CBHW070505200326
41519CB00013B/2723